Management and Applications of Complex Systems

Management and Applications of Complex Systems

Editors

G. Rzevski
The Open University, UK

S. Syngellakis
Wessex Institute, UK

WIT*PRESS* Southampton, Boston

Editors:

G. Rzevski
The Open University, UK

S. Syngellakis
Wessex Institute, UK

Published by

WIT Press
Ashurst Lodge, Ashurst, Southampton, SO40 7AA, UK
Tel: 44 (0) 238 029 3223; Fax: 44 (0) 238 029 2853
E-Mail: witpress@witpress.com
http://www.witpress.com

For USA, Canada and Mexico

Computational Mechanics International Inc
25 Bridge Street, Billerica, MA 01821, USA
Tel: 978 667 5841; Fax: 978 667 7582
E-Mail: infousa@witpress.com
http://www.witpress.com

British Library Cataloguing-in-Publication Data

A Catalogue record for this book is available
from the British Library

Library of Congress Catalog Card Number: 2019932595

ISBN: 978-1-78466-367-4
eISBN: 978-1-78466-368-1

The texts of the papers in this volume were set individually by the authors or under their supervision. Only minor corrections to the text may have been carried out by the publisher.

The material contained herein is reprinted from issues of the International Journal of Design & Nature and Ecodynamics, Volume 13, Numbers 2 and 3, published by WIT Press.

Preface

This volume comprises papers on fundamentals and applications of Complexity Science, the science of open systems consisting of large numbers of diverse components engaged in rich interaction. The global behaviour of these systems emerges from the interaction of constituent components and is unpredictable but not random. The key attribute of Complex Systems is the ability to self-organise and adapt to unpredictable changes in their environment.

Papers in this volume address a wide variety of complex issues from physical to social sciences, including topics such as the role of visual elements in the design of complex systems, methods for controlling deforestation, comparison of lean and complex systems, complex production and logistic systems and coevolution of technology, business and society. Readers are offered a rich choice.

Complexity Science is still in the phase of early development and therefore demarcation criteria are not clear. It is not surprising then that there are disagreements on its content. The Editors took the view that good papers should be accepted even if they addressed issues considered by some to be on the margins of Complexity Science.

The papers contained in this volume, as well as others presented at the Wessex Institute conferences are archived in the Institute's elibrary (witpress.com/elibrary) where they are permanently available in open access format to the international community.

The editors are grateful to all the authors for the quality of their contributions as well as their colleagues who helped to review the papers and hence ensure the quality of this volume.

The Editors
2019

Contents

VISUAL ANALYSIS FOR CONCEPTUAL DESIGN OF COMPLEX SYSTEMS

L.D. GITELMAN, O.B. RYZHUK & M.V. KOZHEVNIKOV
Department of Energy and Industrial Management Systems, Ural Federal University, Russia.

ABSTRACT

Managing the development of complex organizational, technological, and socio-cultural systems calls for the construction of conceptual models that would define the properties and links between the elements of the systems and focus the attention of decision makers on the most significant aspects. The authors propose the visual analysis method as a tool for the conceptual design of complex systems. The method is based on the building of a series of visual images that are associated with the purpose sustainable development and holistic representation of the system. At the initial stage, the general structure of a complex system is analyzed and the required degree of detail of its elements is determined. The system is then broken down into elements and a specific visualization tool is selected for each of them so that the properties of the systems that need improving are highlighted. Ultimately, a conceptual design of the system is created that suggests managerial solutions that meet the interests of the stakeholders. The method has been tested in a number of projects that were implemented with the purpose of ensuring the strategic leadership of major manufacturing companies.
Keywords: complex systems, concept, conceptual design, decomposition, model, multi-project, visual analysis, visual language, visualization.

1 INTRODUCTION

A wave of new technologies, their convergence and integration are in the everlasting process of renovation. At the same time, interconnections between technologies and socio-economic and organizational systems are getting wider and deeper. Such systems are super-complex and require customized tools for managing their creation and operation throughout their life-cycle in which the research phase acquires particular importance. It is the scholarly investigation of complex systems that generates new hypotheses, knowledge and concepts.

As shown in [1], amid uncertainly and risk, the key success factor in forecasting the development of systems lies in anticipatory management. The adoption of anticipatory management practicies is a radically new project that constitutes a serious challenge for business as well as for education and science. Essentially, the question is about conducting an overhaul of the paradigm of development models and forging links within the smart partnership of science, education, business, and government.

The multidisciplinary nature of the smart partnership calls for constructing a common language of concepts and new communication formats [2]. Using a combination of various language systems provides a fuller set of means for conveying the meaning, compared to rather scanty capabilities of the semiotic systems of spoken and written language [3]. That's why interdisciplinary communication has generated a fundamentally new semiotic system that is called visual - verbal language. Along with other actively developing methods, there is an opportunity for the accelerated development and use of visual language for the purpose of anticipatory management. Therefore, it becomes appropriate to use visualization, which means any technique for representing data as a graphical image delivering a holistic representation of a complex subject matter, for addressing the tasks of anticipatory management at the current stage of the development of science and advanced practices in this field [4, 5].

DOI: 10.2495/DNE-V13-N2-166-175

Anticipatory management puts forth new content requirements for research of complex systems [1]. The requirements pertain to the adoption of new instruments of representation of the future, design of its individual elements, first and foremost of promising markets, technologies and properties of new products and services. The issue of creating value for customers interacting with the system comes to the forefront. In the course of design, it is necessary to lay out changes in technological processes, management systems, business models, and competences. The research results in a sequence of transformations that one needs either to undertake right now or to ensure the preparedness for change when a realistic scenario of the future is identified.

Systems engineering is an advanced technique of complex systems analysis [6,7]. Its toolkit, including new research areas, has seen intense development [8–11]. Conceptual design is a starting point in the use of this technique [1]. The novelty of the proposed solution lies in the application of visualization tools for analysis of conceptual models and keeping a research log and recording research results by means of visual language. The authors suggest that visual analysis that contains the principle of the logical construction of visual images should be used as the prime instrument of conceptual design.

2 TRANSFORMATION OF THE LANGUAGE OF CONCEPTUAL DESIGN

The phenomenon of visualization has only recently started to be given attention in science and practice, and so has visual language that emerged as a new mode of communication on its basis. Images and visual artifacts are not just add-ons to verbal texts, mere transmitters of information, or means of communication: they have become an elementary mode for the construction, maintenance, and transformation of meaning [12].

The facticity of visuals is further enhanced by the fact that argumentation disappears. Aspects of visual artifacts are more difficult to ignore compared to aspects of verbal texts. Visuals communicate with an immediacy of reception and a memorable impression of the essence of the message—an essence that, in linear verbal texts, has to be assembled piece by piece over time. It is for these reasons that a consideration of the visual mode of discourse and meaning construction will augment the understanding of contemporary organizational practices, and that it therefore has to play a prominent role in organizational and management research [12,13].

Visual language can be defined as the tight integration of words and visual elements and as having characteristics that distinguish it from natural languages as a separate communication tool as well as a distinctive subject of research. The language was created as the merger of component visual vocabularies from such widely separate domains as engineering diagramming, visual metaphors, cartoons, quantitative graphs and charts, illustration and photography, mechanical drawing, film narrative, and animation techniques [14].

Along with examinations of the meaning and research of tools of visual language, the science of visualization has been given scholarly attention [15]. Coupled with digital tools, visual language could eliminate the main barriers to thinking and the transfer of big complex ideas among people. It effectively solves the problem of the representation and transfer of mental models, which is particularly relevant in the case of poorly structured problems.

Factors driving the development of visual representation techniques are shown in Table 1.

Consequently, the use of visualization enriches the methodology of designing conceptual models of complex systems and boils down to choosing suitable visualization instruments that are tailored to treat ideas as systemic wholes.

Table 1: Drivers of development of visual representation techniques.

Changes in data volumes	
Exponential growth of information.	Problem of studying huge volumes of intricately structured data
Irregular data volume	Need to solve problems with excessive or incomplete information captured in a trend or range format
New user tasks	Unconventional filtering and detalization requests
Growing complexity of systems	
Complex structure tasks	Demand for supplementary instruments for task optimization and solution identification due to limitations of smart systems
Demand for modelling	Visualization of processes and systems for the purpose of research and improvement
Rapid feedback to changes	Management optimization on the basis of system status indicators
Growing computational capacity	
Growing capacity of communications equipment	Growing computational capacity of CPUs, monitors. Higher bandwidth of networks
New generation of software	Mathematic methods and algorithms of visualization. Advancement of graphics software, including those incorporating VR&AR
Controlled interaction	Dynamic feedback between control and controlled systems through visual interface
Transformation of communication modes	
Natural language loses its exclusive role as transmitter of meaning	Growth of verbal – graphical means of information transfer
Information overload	An avalanche of messages with complex concept structures
Interdisciplinary teams	Demand for unified conceptual vocabulary for experts in various domains and from different countries

3 HANDS-ON PRACTICE WITH VISUALIZATION IN CONCEPTUAL DESIGN

Hands-on practice with visualization is delivered with the application of conceptual design techniques that reveal various aspects of the operation of systems. The result of the experiment is a conceptual model of a complex system that includes a system of ideas and relationships among them [6].

In Table 2, the authors sum up visualization tools that are utilized for describing the key properties of complex systems.

When constructing a conceptual model of a complex system, one encounters the following typical problems:

1. identifying classes and groups of elements within the complex system and establishing links between them;

Table 2: Visualization tools for describing key properties of complex systems.

Attribute	Visualization tool
Openness that lies at the heart of active interaction with environment	**Rich picture** Sketches, drawings, photographs, symbols, titles that depict a situation or an issue from the perspective of a person or a group of people. It covers people, systems, processes, interfaces, information, infrastructure units, factors, emotions.
Diversity, partial autonomy and links between agents	**Communication diagram** Communication Diagram models the objects or parts of a system, the interactions (or messages) between them, and the sequence in which these interactions occur. A communication diagram is a combination of information, its sequence and application options. It describes both the static and dynamic behavior of the system. Communication diagrams use a free-form arrangement of objects and their interactions within a system.
Agent self-organization and absence of centralized control over agents	**Semantic network** It represents a subject area as a directed graph with nodes and arcs that link nodes. Networks can cover concepts, events, properties, processes. A network reflects the semantics of the subject matter in the form of concepts and relationships. The items of a semantic network are placed into simple shapes and connected with labelled arrows.
Indeterminate behavior of agents that implies freedom of choice and, at least, several pathways to results	**Data-flow diagram** A data flow diagram (DFD) is a graphical representation of the "flow" of data through an information system, modelling its process aspects. It shows what kind of information will be input to and output from the system, how the data will advance through the system, and where the data will be stored. DFD are often used to visualize a holistic vision of the system without going deep into detail.
Emerging properties of a self-organizing system that are absent in agents that constitute it	**System dynamics (simulation)** The systems dynamics approach models the nonlinear behavior of complex systems over time using stocks, flows, feedback loops, table functions and time delays. It reflects the structure of a system and the timeframe of interactions among its components. Mathematical modelling methods are employed.

2. deciding on the required and sufficient level of detail in the complex system representation;
3. choosing the language and format of element representation to express the content of the model.

The above difficulties make it impossible to use modelling software at the initial stage – that of constructing a conceptual model. This produces the need to compare the research toolkit, the representation format and the level of complexity in the system with human capabilities. In order to address the task one needs to use to

- use instruments that make it easier to systemically organize elements and links;
- present a super-complex system in an overview;
- pick an all-encompassing form of representing the content of the conceptual model.

Systemization instruments for elements and links. It is typical of a conceptual model to use the same structures for representing items that are different in nature. Such structures serve as a semantic framework – construct – for various items. Using multi-shaped constructs makes it much easier to identify and systemize system elements. Visualization of the construct helps to quickly classify elements when constructing a system and to set accents for better perception of the information. An example of construct utilization is shown in Fig. 1.

Overview. The intention to collect elements and see a subject as a single system is the main goal of constructing its model [4].

Using an overview simplifies the task of constructing and absorbing a conceptual model. An overview of a complex system is reflected in a well-picked metaphor. Information is arranged graphically, with the nature of elements and interconnections being conveyed through the key characteristics of the metaphor being used. The analysis of the model is guided by the interpretation of the metaphor. The approach enables one to focus on the essential elements and attributes of the system.

The list of metaphors that are used for organizing the elements of complex systems includes "tree", "metro map", "iceberg", "ruler" [16–19]. Some metaphors also serve as the basic constructs. For example, the "tree" metaphor could show the hierarchy in an organization and flows in a financial system. Figure 2 illustrates the use of the "matryoshka doll" metaphor for visualizing the architecture of the "Technology leadership incubator".

Form of expression. Depicting a conceptual model by means of verbal language only strongly narrows down opportunities for its examination and analysis. This is due to the fact that a verbal description of the diversity of its elements and their relationships will be as complex as the system itself. Other forms, too, have serious limitations because they use a certain vocabulary. It is possible to say that replacing one semiotic system with another affects the interpretation of the conceptual model and does not deliver the level of reflection that is needed in research. It is therefore necessary to use a language that offers a diversity of forms of expression: formulas, graphics, etc. An example of a combination of visual and verbal forms is given in Fig. 3.

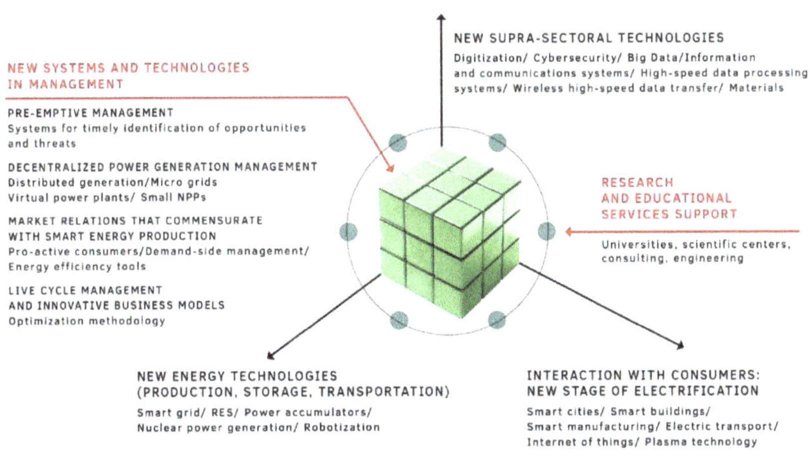

Figure 1: Construct utilization in the case of power engineering.

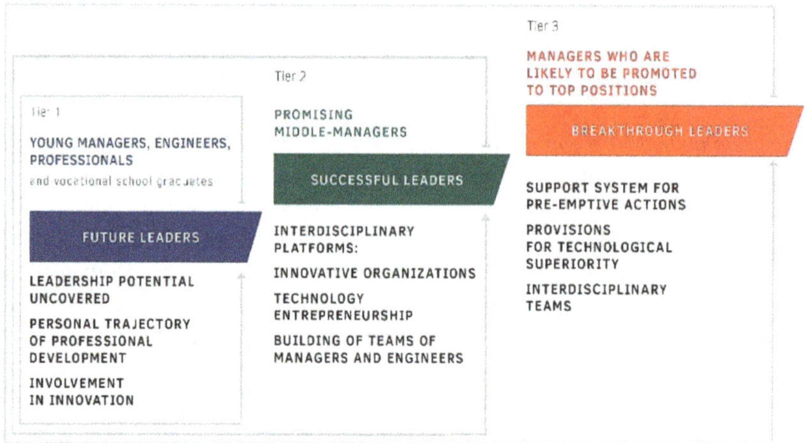

Figure 2: Metaphor utilization for producing an overview of the architecture of the Technology Leadership Incubator.

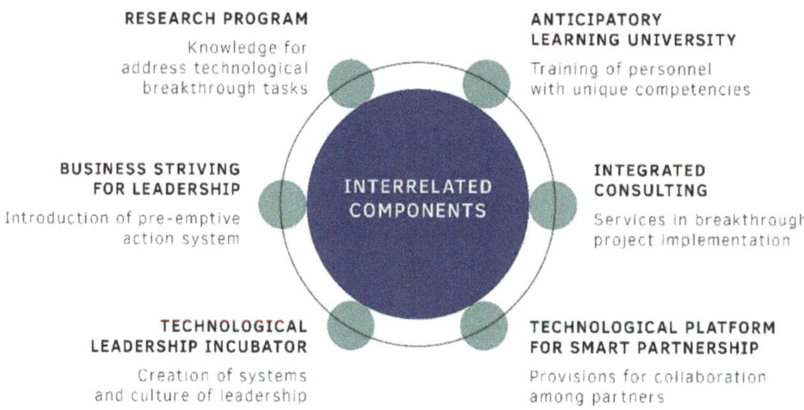

Figure 3: Combination of visual and verbal forms for graphic representation of the multi-project "New leaders for technological breakthrough".

The visualization method reveals the attributes that have the biggest importance for the purpose of research. It illustrates the features of the inner structure and the principle of relationships among the elements of a complex system, thus helping to propose ways of augmenting it.

4 RESEARCH RESULTS

The proposed method was tested as part of educational projects tailored for master's students and corporate customers of the Research and Educational Centre for Interdisciplinary Investigations and Educational Programs (REC ENGEC) at Ural Federal University in 2017–2018.

During organizational activity games, practitioners and university faculty members attended discussions of projects that aimed to address complex tasks of industrial modernization. Around 250 persons took part in special business games in 2017–2018. Some of the visual solutions to conceptual projects that were presented at the games are given below.

Figure 4 shows the project of an online sales platform for the Novolipetsk Steel Works (NLMK Group), a leading vertically integrated steel manufacturer with assets in the US, Russia, France, Belgium, and Denmark. The integrated online sales platform show in the figure would provide for prompt interaction with end consumers, decrease delivery times, improve the flexibility and efficiency of capacity utilization thanks to control over the entire value chain.

Figure 5 shows a project aimed at the expansion of the utility market for T-Plus, the largest privately owned energy company in Russia. Today, the company only produces and sells heat

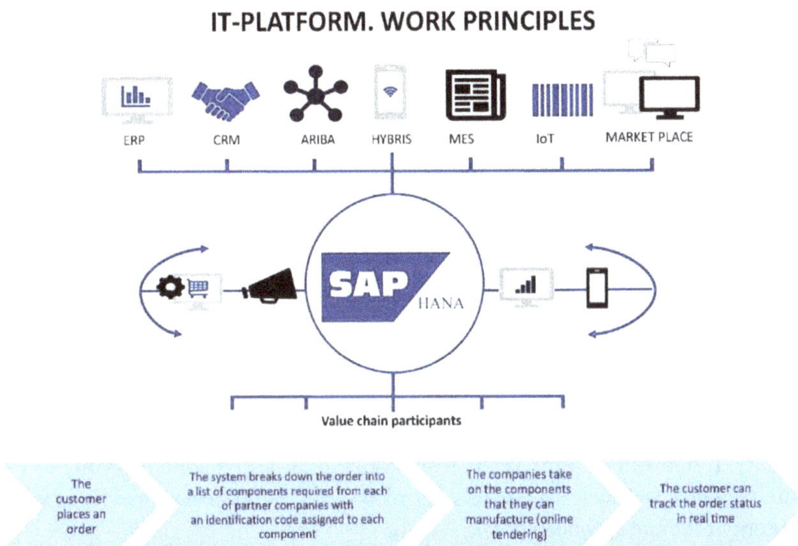

Figure 4: Online sales platform for NLMK Group

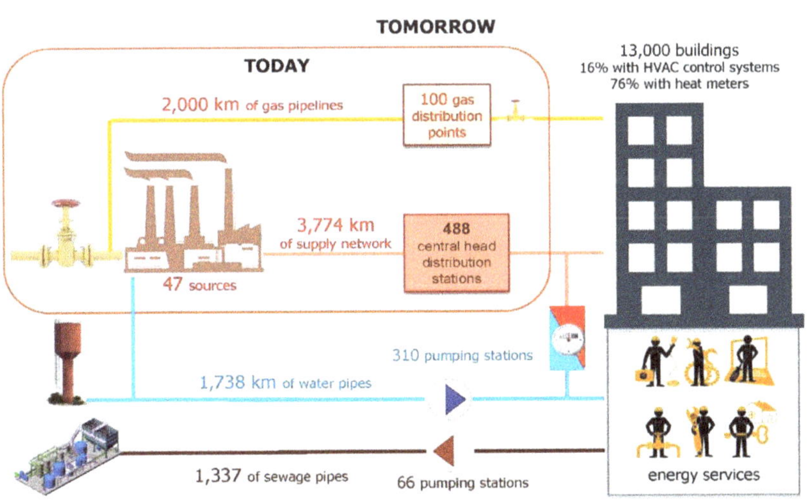

Figure 5: Expansion of T Plus utility market

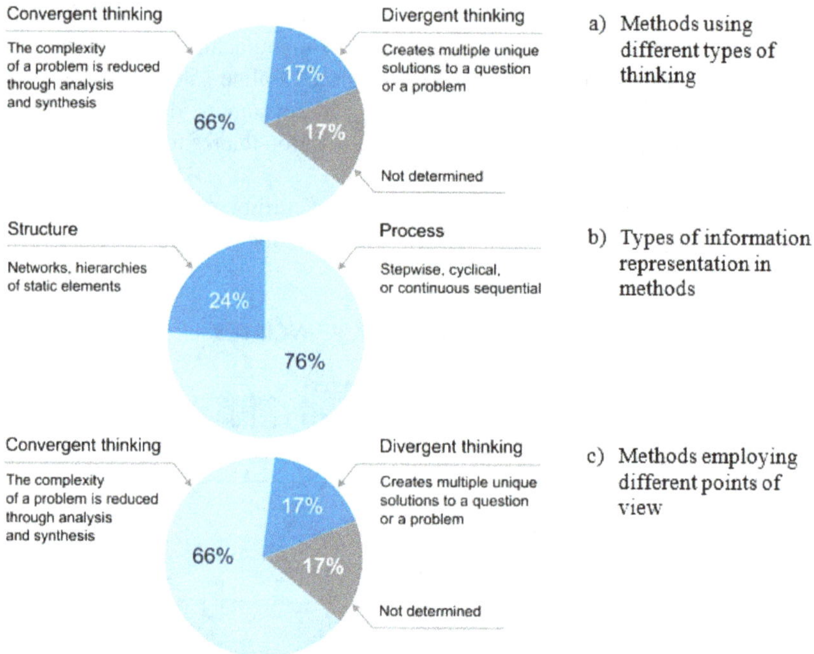

Figure 6: Distribution of visualization methods.

and electricity. It plans to acquire assets from companies with a similar profile to be able to provide maintenance services for HVAC control systems and house connections with the use of smart energy efficient technologies.

The results of practical application are backed with data produced by studies into the cognitive attributes of visualization methods [5]. The attributes could be divided into three groups: type of thinking; type of representation; and point of view. The percentage distribution of the attributes is found in Fig. 6.

5 CONCLUSION

Against the backdrop of accelerated digitization, convergence of markets and intrinsically difference systems, structures and development factors, visualization plays the role of an integrator that makes up for the absence of a unified conceptual framework and services as a "lighthouse", providing a clear map of priorities and targets for decision makers. It is therefore natural that the subject of visualization has started to attract growing attention both in science and in the areas of its application.

The novelty of this article lies in the fact that it investigates opportunities for visualization at the stages of complex systems design when no information is available yet about the ultimate parameters of the system, its interface and empirical characteristics. In this context, methods of visual analysis make it possible to test a hypothesis in terms of its overview and perception of the new system and to start a discussion in the multidisciplinary community of experts.

Professionals with visualization skills are needed for the successful adoption of visual language in complex systems management at the initial stage of the systems' lifecycle. At the moment, no such specialists can be found. This highlights the urgency of developing educational programs that would nurture such competences. This would turn visual language into a natural form of interdisciplinary communication.

ACKNOWLEDGEMENTS

The work was supported by Act 211 of the Government of the Russian Federation, contract No 02.A03.21.0006.

REFERENCES

[1] Gitelman, L.D., Sandler, D.G., Gavrilova, T.B. & Kozhevnikov, M.V., Complex systems management competency for technology modernization. *International Journal of Design & Nature and Ecodynamics*, **12(4)**, pp. 525–537, 2017. https://doi.org/10.2495/dne-v12-n4-525-537

[2] Gitelman, L.D., Sandler, D.G., Kozhevnikov, M.V. & Tretyakov, V.S., Technology platform as a tool for transformation of university science and education activities, [in Russian]. *University Management: Practice and Analysis*, **4(98)**, pp. 31–42, 2015.

[3] Kress, G., *Multimodality: A Social Semiotic Approach to Contemporary Communication*, London: Routledge, pp. 212, 2010.

[4] Shneiderman, B., The eyes have it: a task by data type taxonomy for information visualizations. *Proceedings of 1996 IEEE Conf. on Visual Languages (IV03)*, pp. 336–343, 1996. https://doi.org/10.1109/vl.1996.545307

[5] Lengler, R. & Eppler, M.J., Towards a periodic table of visualization methods for management. *Proceedings of the IASTED International Conference on Graphics and Visualization in Engineering*, pp. 83–88, 2007.

[6] Rzevski, G. & Skobelev, P., *Managing Complexity*, Southampton, Boston: WIT Press, 2014, 198 p.

[7] Gavrilova, T.B., Gitelman, L.D. & Kozhevnikov, M.V., *System Engineering for Managers* [in Russian]. Moscow: Economics, 2017, 188 p.

[8] Woods, D.D., Four concepts for resilience and the implications for the future of resilience engineering. *Reliability Engineering and System Safety*, **141**, pp. 5–9, 2015. https://doi.org/10.1016/j.ress.2015.03.018

[9] Gillani, S.M., Qadri, S. & Fahad M., Customer oriented requirement engineering by using scrum methodology. *International Journal of Natural and Engineering Sciences*, **8(3)**, pp. 7–11, 2014.

[10] Barry, P. & Dove, R., Combating uncertainty in the work flow of systems engineering projects. *INCOSE International Symposium*, **23(1)**, pp. 368–382, 2013.

[11] Barelkowski, R., Design process as complex system. *International Journal of Design & Nature and Ecodynamics*, **13(1)**, pp. 46–59, 2018. https://doi.org/10.2495/dne-v13-n1-46-59

[12] Meyer, R.E., Höllere, M.A., Jancsary, D. & van Leeuwen, T., The visual dimension in organizing, organization, and organization research: Core ideas, current developments, and promising avenues. *The Academy of Management Annals*, **7(1)**, pp. 487–553, 2013. https://doi.org/10.1080/19416520.2013.781867

[13] Vittikh, V.A., *Introduction to the Intersubjective Management Theory* [in Russian], Samara: Samara Scientific Center of the Russian Academy of Sciences, 2013, 64 p.

[14] Horn, R.E., *Visual Language: Global Communication for the 21st Century*, Bainbridge, WA: MacroVU Press, 1998, 270 p.

[15] Willems, K., The Periodic Table of Data Science. The 150+ companies, resources, and tools that define the data science industry, available at https://www.datacamp.com/community/blog/data-science-periodic-table?utm_campaign=crowdfire&utm_content=crowdfire&utm_medium=social&utm_source=instagram#4542244293-in#1519033028315 (accessed 10 March 2018)

[16] Averbukh, V., Visualization metaphors. *Programming and Computer Software*, **27(5)**, pp. 227–237, 2001.
https://doi.org/10.1023/a:1012333025189

[17] Risch, J.S., On the role of metaphor in information visualization, available at https://arxiv.org/pdf/0809.0884.pdf (accessed 10 March 2018)

[18] Lima, M., Visualization metaphors: old & new, available at https://www.interaliamag.org/articles/visualization-metaphors-old-new/ (accessed 10 March 2018).

[19] Ziemkiewicz, C. & Kosara, R., The shaping of information by visual metaphors. *IEEE Transactions on Visualization and Computer Graphics*, **14(6)**, pp. 1269–1276, 2008.
https://doi.org/10.1109/tvcg.2008.171

APPLICATION OF POWER LINE COMMUNICATION TECHNOLOGY IN STREET LIGHTING CONTROL

TEKLA D. FARKAS[1], TAMÁS KIRÁLY[2], TAMÁS PARDY[1], TOOMAS RANG[1] & GALINA RANG[1]
[1]Thomas Johann Seebeck Department of Electronics, Tallinn University of Technology, Estonia.
[2]Flashnet Hungary Kft, Hungary.

ABSTRACT

Rapidly increasing usage of telecommunication systems causes new transmission technologies and networks to emerge. Not only the efficiency, reliability and accessibility of the network are important, but also the economic issues. One cost-effective solution could be power line communication (PLC) technology, which transmits data using the existing electricity infrastructure. The application of this communication technique is an attractive and innovative solution for the realization of smart cities and smart homes. With intelligent control networks, energy savings can be optimized and the operating as well as maintenance costs can be reduced. Since outdoor lighting systems are the major consumers of electricity, to create a modern, energy-efficient city, intelligent street lighting control is needed. This paper provides an overview of power line communication principles including the theoretical background of data communication, modulation techniques, channel access methods, protocols, disturbances and noises. Furthermore, in order to highlight the benefits of a PLC-based street lighting control system, a pilot project will be presented.

Keywords: intelligent lighting control, modulation techniques, Power line communication, smart city, smart grid.

1 INTRODUCTION

Power line communication (PLC) is a very attractive technology, which can be used for many applications, such as home automation systems, CCTV cameras, pollution sensors, noise detectors, traffic density sensors, advertising backlit panels, solar applications, energy measurements, street lighting control or vehicle communication networks [1].

Energy usage of street lighting accounts for a significant portion of the total energy consumption of a city [2]. In order to reduce energy consumption, CO2 emission, light pollution, to improve public safety and enhance public service quality, implementation of an intelligent light control system is essential [1]. PLC is a natural choice for supporting intelligent lighting control systems in order to ensure the optimal illumination level under any operational circumstances and to reduce operational costs.

A PLC network can also be the backbone communication infrastructure for a smart grid, which enables reliable, real-time measurements, monitoring and control [3]. The implementation of this management solution is cost-effective, since the communication is done over the existing power lines and therefore no additional infrastructure is needed. Due to this fact, PLC technology can compete with wireless technology and can be a communication medium in a future smart grid [4].

There are three basic network elements for the realization of a PLC system: base station, modem and repeater. PLC technology provides a link between the subscribers and their immediate service provider. The application of PLC in the power supply grid can be considered as a communication solution for the access network. To connect the access network to a wide area network (WAN), a PLC base station is used. To make communication possible over power supply grids, PLC modems are needed between the communication devices and power lines, which convert data signals into a suitable form for both directions of transmission. A PLC modem has two interfaces: on the user side, it provides Ethernet and USB network

DOI: 10.2495/DNE-V13-N2-176-186

interfaces and in order to connect the PLC modem to the power line, PLC signal injection should be realized by inductive or conductive coupling. The PLC modem carries out the function of the physical layer (PHY) and the data-link layer (DLL, including both of the sublayers, the media access control (MAC) and the logical link control (LLC)). In case of longer network distances, repeaters are used to ensure sufficient data rate. The repeaters divide the PLC access network into network segments. In order to separate the segments, different frequency bands or time slots are used. With segmentation, signal power and electromagnetic radiation can be reduced, but network capacity decreases, because wider frequency bands have to be used to avoid the interference between the segments [5, 6].

Gateways are usually used for better network efficiency, dividing the PLC access network into subnetworks. Gateways not only forward the data between the networks, but can also interconnect networks, which use different network protocols by performing the required protocol conversions. The drawback of this setup is that repeaters and gateways are expensive elements and because of the signal conversion, propagation delay can occur. It is also of note that the frequency ranges are limited in the case of PLC, and losses on the medium can negatively influence power transmission. Various noise sources (different electrical devices connected to the power supply grids with the same electromagnetic spectrum) can also cause disturbances in the PLC service [5, 6]. In this paper, we provide a short overview about the theory, principles and basic features of PLC systems. Furthermore, to highlight the attractiveness of this technology, a PLC based street lighting control system will be introduced.

2 THEORETICAL BACKGROUND AND BASIC FEATURES OF PLC SYSTEMS

2.1 Frequency ranges for PLC

There are two frequency ranges for PLC: narrowband and broadband. Defined by CENELEC, narrowband PLC (NB-PLC) networks operate between 3–148.5 kHz in Europe. This frequency range is divided into four bands: A, B, C and D. Band A [3–95 kHz] is restricted to the energy suppliers, frequency bands B [95–125 kHz], C [125–140 kHz], D [140–148.5 kHz] can only be used by consumers. Typically, NB-PLC systems operate between 35–90 kHz in the CENELEC A band, while broadband PLC (BB-PLC) networks use the 3–34 MHz frequency band [5, 7].

The NB-PLC systems can provide data rates up to few thousand bps, while the BB-PLC systems have reached few hundred Mbps [8]. These higher data rates can enable high-speed data transmission, multiple voice connections and transfer of video signals as well. The main differences between narrowband and broadband PLC are in frequency range, information capability, power and attenuation [5, 7].

2.2 Modulation techniques

Modulation techniques are methods, which are used to code analogue data into digital format. There are various types of digital modulation schemes depending on the type of the signal and the application [9]. The most common types will be described below.

1. Amplitude Shift Keying (ASK): in this case, the carrier signal is modulated with a binary message (0 or 1). The carrier signal is transmitted, if the signal value is 1, but when this value is 0, the carrier signal will not be transmitted. The amplitude shift keying (ASK) technique is sensitive to disturbances and it has not constant envelope,

which makes power amplification more difficult. The first NB-PLC systems were realized by this modulation technique [5, 10].

2. Frequency Shift Keying (FSK), Binary Frequency Shift Keying (BFSK): digital information is transmitted through discrete frequency changes of the carrier wave. The two different digital values of 0 and 1 represent two different frequencies [10]. This modulation technique can be used for narrowband and broadband communications depending on the separation of the two frequencies [5]. Although it is a cost-effective solution, it is not bandwidth-efficient and requires a complex receiver design.
3. Phase Shift Keying (PSK), Binary Phase Shift Keying (BPSK), Quadrature Phase Shift Keying (QPSK): when the carrier wave phase is discretely varied, the modulation scheme is called phase shift keying (PSK). In the case of BPSK, two different digital values (0 and 1) are used and when a binary change occurs in the serial binary data, the carrier wave shifts 180°. For QPSK, bit pairs are used. Each bit pair represents different carrier phase shifts. The bit pairs and phase degrees are the following: 00–45°, 01–135°, 11–225°, 10–315°, where for instance a phase change of 45° means 00. The amount of the information associates with a phase or the number of the discrete phases that the carrier wave contains. In case of ASK, BFSK and BPSK, the baud rate is 1 bit/symbol, while QPSK can transmit 2 bits/symbol [11, 12].
4. Quadrature Amplitude Modulation (QAM): quadrature amplitude modulation (QAM) is the combination of amplitude and phase modulation. For instance 16-QAM uses twelve carrier phases plus three amplitude levels to transmit 4 bits per symbol [11, 12].
5. Orthogonal Frequency Division Multiplexing (OFDM): orthogonal frequency division multiplexing (OFDM) is a multicarrier modulation technique. In contrast with single-carrier, the data is not transmitted by only one carrier, but they are shared among several carriers and are transmitted simultaneously. The single high-rate data stream is subdivided into N low-rate parallel substreams. Each substream is modulated over a separate carrier signal subcarrier, moreover each subcarrier can be modulated differently (BPSK, QPSK etc.). The subcarriers are not separated but overlapped and they are orthogonal, which means that the peak of one subcarrier occurs when other subcarriers are zero [5, 13, 14]. OFDM has several advantages. It is robust in a multipath propagation environment, more resistant against fading, tolerant to delay spread and it can minimize the effect of intersymbol interference (ISI) [5]. Since N different subcarriers are used, the symbol duration of a signal is N times longer than the signal duration of the original signal. To avoid interference, a guard interval (cyclic prefix) is used between the signals. The cyclic prefix is a copied portion of the end of a symbol, which is put at the head of the symbol.

2.3 Spread-spectrum techniques

The spread-spectrum method is a technique in which a pseudo noise code, independent from information data, is used as a modulation waveform. It results in greater bandwidth than the signal information bandwidth. The big advantage of these techniques is that the broadband, noise-like signals are more resistant to interference, noise and jamming, and it is harder to detect, intercept and demodulate them than narrowband signals [5, 15]. Commonly used techniques are listed below:

1. Direct Sequence Spread Spectrum (DSSS): In this method, the binary data (bits) are logically compared to a string of pseudo noise (PN) code bits (chips). The chip rate

$(R_c = 1/T_c)$ is much higher than the data rate $(R_s = 1/T_s)$. The binary data is multiplied with the PN code to produce the transmitter signal. The effect of the multiplication of the bits and chips is the spread of the baseband bandwidth R_c to the baseband R_s. Afterwards, usually BPSK or PSK modulations are used to transmit the signal to the high frequency. The receiver compares the spread spectrum signal with the same chip sequence used in the transmitter and then, a demodulator is applied to reproduce the transmitted data [15].

2. Frequency Hopping Spread Spectrum (FHSS): This method spreads narrowband transmitted signals, hopping one frequency to the next. The transmission frequencies change according to the pseudo noise sequence. Compared to the above-mentioned spread-spectrum techniques, DSSS can provide higher data rate than FHSS, since in order to increase the data rate at DSSS, only the clock rate has to be increased, while at FHSS a wider bandwidth is needed. DSSS transmission is continuous and broadband, while FHSS transmission is discontinuous and narrowband. FHSS is more susceptible to narrowband noise and interference, than DSSS [5, 15].

2.4 Channel Access Method (CAM):

Channel access methods are used to allow individual users to use the same fixed bandwidth. There are three basic access schemes depending on which channel is separated, but their combination can also be used [13]:

1. Frequency-division multiple access (FDMA): in an FDMA scheme, the available system bandwidth is divided into non-overlapping bands. In order to reduce the interference, there is a guard band between the adjacent frequencies. The frequency bands are assigned among the users, so that no other user can use the same band. The frequency band can be reassigned, when the user terminates the service [13].
2. Time division multiple access (TDMA): TDMA allows the users to use the same band by dividing the signal into different time shots. To avoid that data from adjacent time slots interfere, there is a protection intervals (guard period) between the time slots. A certain number of timeslots forms a TDMA frame. When disturbances occur, just the erroneous, but segmented data packets have to be retransmitted, therefore this scheme is widely used in PLC technology [5, 13].
3. Code division multiple access (CDMA): the CDMA access method transmits the data according to a special coding scheme. This method enables to send information simultaneously over a single communication channel [13].

2.5 Protocol

In PLC, the medium is listened to by both the PHY level (physical carrier sense) and the MAC layer level (virtual carrier sense). In order to avoid collision by different users, the channel is listened to before sending the data. If the medium is free for a DCF inter-frame space (DIFS) interval, transmission is permitted. If the frame is successfully received, the receiver sends an acknowledgment frame (ACK) to notify the source of successful reception. The time between the reception of a frame and the transmission of the ACK is called short inter-frame space (SIFS). If the channel is used, the sender must wait a random period of time before attempting to access the medium again (back off time). In case of virtual carrier sense mechanism, first the sender transmits an RTS (request-to-send) packet, which indicates the transmission time required for the data and the ACK packet. Receiving the RTS packet, the destination station sends back a CTS (clear-to-send) packet after SIFS time. Implementation of the RTS/CTS mechanism can help avoid collisions, since each station in the range of the receiver is informed about the transmission [5].

2.6 Coupling

There are two different methods, inductive and capacitive coupling, to connect PLC devices to the transmission medium [16, 17]. The basic function of the coupling circuit in a PLC modem ensures the maximum power for both transmission and reception modes. It protects devices from transient components and rejects noise signals. Inductive coupling is noisier than capacitive coupling, but it can reduce attenuation by 10 to 15 dB for some frequencies [16, 18]. It can also be an advantage that no physical connection to the grid has to be made, because the inductive couplers are electrically insulated from the power cables. In case of inductive coupling, a ring-shaped ferrite is clamped around the cable in order to inject the communication signal. The inductive coupler can be considered a transformer, which has a modem-side primary winding, and a power cable side secondary winding. In this way, the signal is directly induced through the coupler, to and from the power cable. Capacitive coupling is the preferred method at coupling points with relatively high impedance (>20 Ω) and when inductive coupling is not possible due to high current ratings, or the geometry of the couplers cannot allow the access due to too small inner diameter of the ferrite [16, 17].

2.7 Power supply grid

Realization of the PLC network depends on the power supply grid levels. There are three power supply grid levels, high-, medium- and low-voltage level. The high-voltage grid (110 kV or above) is the backbone of the electricity supply system. It is a part of the transmission stage and connects power stations with large supply regions or big consumers. It enables long-distance transmission using high voltage (110 kV, 220 kV, 330 kV, 400 kV, 750 kV) in order to reduce line losses. Power is usually transmitted through overhead power lines, but in urban areas or sensitive regions underground power transmission is used. The medium-voltage grid (usually 10–30 kV) supplies urban and rural areas. Overhead and underground power lines are also used. MV grids usually have ring topology. Low-voltage grid (400 V) supplies households and small industrial clients. Home electrical installations belong to the low-voltage level, but are owned by the users [19]. PLC systems are primarily used in medium- or low voltage grids. The attenuation is determined by the cable and coupling losses. The attenuation is lower and less frequency selective on the medium-voltage grid, than on the low-voltage grid. On the other hand, attenuation shows frequency-dependence; the coupling loss decreases with increasing frequency. From a certain frequency, the cable loss becomes dominant, therefore the attenuation increases again. An underground MV power grid is less influenced by noises, compared with LV power grid [5, 19].

2.8 Disturbance, noises

Power supply grids are not primarily designed for communication. The transmission channel is characterized by different types of noises, frequency-dependent attenuation, changing impedance and fading. The physical properties of electrical wiring influence the transmission of the data. Various electronic devices (resistive, inductive and capacitive elements) connected or disconnected to or from the grid constantly modify the inductive and capacitive characteristics of the grid. On the other hand, PLC systems produce electromagnetic radiation, which can cause disturbances in other services operating in the same frequency range [20].

2.9 Electromagnetic compatibility (EMC)

Electromagnetic compatibility refers to emissions and susceptibility of different systems, which use the same frequency band. The classification of EMC disturbances according to the spectrum can be seen on. In case of PLC systems, the main source of EMC is the radiated and conducted emission from the asymmetric cables in LV and MV grids [5, 21].

2.10 Noise

Noise can be categorized into the following classes [5]:

1. Coloured background noise: this noise is a slow, time-variant random disturbance. The power spectral density decreases with increasing frequency.
2. Narrowband noise: this noise is mainly caused by broadcasting and wireless communication systems. The radio signals of these can impair the PLC channel. Narrowband noise mostly has a sinusoidal waveform.
3. Periodic impulsive noise, asynchronous to the main frequency: this noise is mainly caused by switching power supplies. Its frequency range is from 50 kHz to 2 MHz with a form of impulses which has a repetition rate between 50 kHz and 200 kHz. This type of noise results in discrete lines in the spectrum.
4. Periodic impulsive noise, synchronous to the main frequency: this type of noise is mainly caused by the power supply, which operates synchronously with the main frequency. The repetition rate of this impulse is synchronous with the frequency mains or its double.
5. Asynchronous impulsive noise: this noise is usually caused by switching transients. Duration of impulses is from a few microseconds to milliseconds with arbitrary inter-arrival times. The asynchronous impulsive noise has high power spectral density (greater than 50 dB), thus this class of noise causes most PLC system errors [8].

3 PILOT PROJECT: INTELLIGENT LIGHTING CONTROL FOR LIGHTING NETWORK SYSTEM

3.1 Experimental setup

3.1.1 Overview

Intelligent control of street lighting systems is a vital part of smart city projects. It can offer several benefits and reliable solutions for a city: it can help save energy by dimming light sources, enhance safety by real time monitoring and analysis of the network and reduce maintenance costs by intelligent optimization [1]. In the following, a pilot project will be introduced in detail, which is operated in Pilisborosjenő, in Hungary [22]. The system architecture consists of three levels: lamp-level, remote terminal unit level and software control level. The control and monitor of the lamps are realized with InteliLIGHT Streetlight Management System. 27 street lighting fixtures are controlled, of which 11 are high-pressure sodium lamps (4 pieces of 100 W light bulbs and 7 pieces of 70 W light bulbs), 13 are compact fluorescent lamps (36 W), 2 are LED lamps (18 W, 36 W) and 1 is a metal-halide lamp (400 W). Each lamp has its own ON/OFF switching and/or dimming function, providing the possibility to control them individually. The controllers are mounted on street lighting fixtures, while the communication devices (electric network analyser, data concentrator, communication router and modem) are placed inside the power supply cabinet. PLC is used between the controller and communication devices, and IP-based communication is used for data transfer between the controllers and the management software [1]. Figure 1. shows the location of the lights, including a table that contains the light bulb types and controllers. Figure 2. shows the hardware used for control and analysis.

3.1.2 Controllers

FPM-150 and FPM-400 (up to 150 W and 400 W respectively) type electromagnetic ballast controllers were used for HPS and MH lamps. The controllers switch the lights on/off and control the brightness setting (dimming). They are also responsible for collecting information about the system status (failures, faults etc.) and features (voltage, current etc.). FPX-220 on/

Color	Type of bulb	Type of controller
	HPS	FPM-150
	LED	FPE-100
	HPS	FPM-400
	Compact	FPM-400
	MH	FPX-220
	Compact	FPM-150

Figure 1: Locations of the lighting fixtures, types of bulbs and controllers

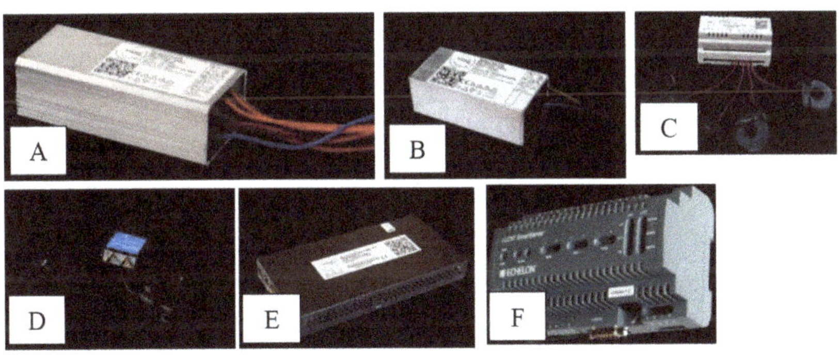

Figure 2: FPM-400 (A), FPX-220 (B), FNM-110 (C), Teltonica router and FRB-100 modem (D&E), FPC-200 (F)

off electronic controller can be used for any types of lamps, but it does not provide dimming control function. Since constant illumination was needed for the floodlighting of the church, FPX-220 was a good choice for the remote management of the luminaries. The controllers use the CENELEC C band and LonWorks, peer-to peer protocol. It uses a narrowband spread-spectrum modulation scheme and CSMA in order to avoid data collisions [8].

3.1.3 Electric network analyser (FNM-110)

Electric network analyser can enable measurements and analysis of the street lighting grid parameters. It has R, S, T, N current inputs. It can send report about effective voltage and current, frequency, active, reactive and apparent power and power factor.

3.1.4 Data concentrator (FPC-200)

FPC-200 data concentrator is a software interface for configuration, programming, system integration and failure alarming. It receives the data from the components and converts the information using LonWorks protocol. After converting the data, the information packages are sent to the central processing unit.

3.1.5 Communication router (Teltonica, RUT 104) and modem (FRB-100)

Communication router provides VPN connection between controllers, network analysers, data concentrators and InteliLIGHT Control Software. The router saves data logs and history of the events on the grid and archives them. The FRB100 modem saves and archives data logs and history of the events on the grid and ensures the secure data transmission to the Network Management Software.

4 RESULTS AND DISCUSSION

The lighting fixtures are situated on two different lighting circuits (Phase 1, Phase 3). On Phase 1, mainly high-pressure sodium (HPS) lamps are situated, and ON/OFF electronic controllers are used, while on Phase 3, mainly compact fluorescent lamps are situated, and dimming control is used (see Fig. 1.). Streetlights are controlled according to a predefined schedule, which is based on the experience and demand of the local government and energy supplier. Streetlights are switched ON/OFF at 20 minutes before/after the sunset/sunrise, and the controlling is done at 11 pm and 12 pm to 90% and 80% of the nominal power. To estimate how the active and reactive power can be changed by the control, a one-month long period (01–30.09.2016.) was investigated. On Fig. 3. A, B, the change of the instantaneous and total active power can be seen (only a one-week period is depicted). Since on Phase 1 only ON/OFF electronic controllers are used, the active power was constant (~827 W). In case of Phase 3, where the dimming function is used, the active power was ~1293 W before the control. In the first and second steps of the control process, the value of the active power was ~1238 W and ~1169 W, respectively. This means that the active power decreased with ~9.6% in total; ~4.2% in the first step and ~5.4% in the second step. Table 1 summarizes active power values for phase 1&3.

On Fig. 3. C, D, the change of the instantaneous and total reactive power can be seen for the period of 16.09.2016–23.09.2016. In case of Phase 1, the reactive power was almost constant (~661 VAr). In case of Phase 3 no significant difference can be measured; before the control, the value of the reactive power was 1330 VAr which decreased to ~1328 VAr and ~1296 VAr, respectively (~3% in total). Table 2 summarizes reactive power values for phase 1&3.

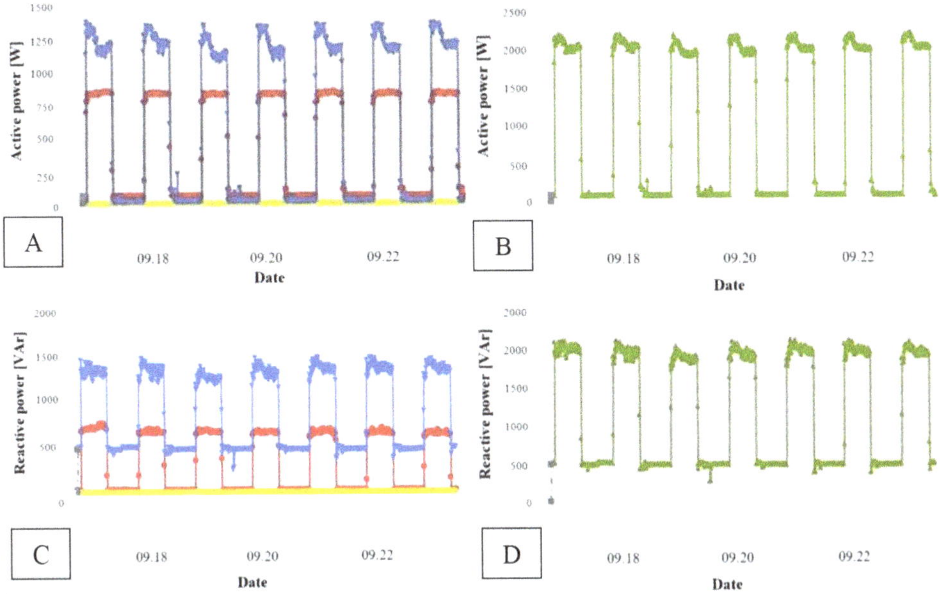

Figure 3: Instantaneous active power (A), Total instantaneous active power (B), Instantaneous reactive power (C), Total instantaneous reactive power (D). Plot colours: red:=Phase 1, blue:=Phase 3.

Table 1: Active power values for Phase 1&3.

	Phase 1 (HPS lamp + ON/OFF control)	Phase 3 (fluorescent lamp + dimming control)
Active power (no control)	~827 W	~1293 W
Active power (step 1)	~827 W	~1238 W
Active power (step 2)	~827 W	~1169 W

Table 2: Reactive power values for Phase 1&3.

	Phase 1 (HPS lamp + ON/OFF control)	Phase 3 (fluorescent lamp + dimming control)
Reactive power (no control)	~661 VAr	~1330 VAr
Reactive power (step 1)	~661 VAr	~1328 VAr
Reactive power (step 2)	~661 VAr	~1296 VAr

These results are in good agreement with the purposes of this study: with intelligent lighting control system, energy consumption can be decreased. It should also be noted that the quality of the power supply grid, such as age and material of the cables, types of cable shielding and insulation, corrosion of cable joints, substantially affects the propagation of the signal. Away from the power supply cabinet the signal strength can drop and the probability and duration of the outage can increase.

5 CONCLUSIONS

In this paper, we reviewed the theoretical background of PLC technology related to smart street lighting control, including frequency ranges, modulation techniques, protocols as well as coupling methods and the characteristic power supply grid. Furthermore, a pilot project demonstrating the benefits of intelligent lighting systems was detailed. The project was carried out in a small town of about 3300 inhabitants with 27 lamps monitored for power consumption as well as controlled through PLC. The pilot study concluded that an up to ~9.6% decrease of active power was achievable by smart lighting control. PLC technology is a low-cost, environment-friendly and flexible communication technology, which is an ideal candidate for the realization of smart grid networks. Future works will be targeted at extending the project to more heavily populated areas, namely cities at least one order of magnitude larger in population. The larger the settlement, the more complex the power grid as well as the more challenging it is to maximise power efficiency while causing minimal disruption to night activities characteristic of larger settlements.

6 REFERENCES

[1] InteliLIGHT® street lighting management system, *FlashNet Srl*, available at https://intelilight.eu/ (accessed 25 September 2016).

[2] How much electricity is used for lighting in the United States?, U.S. Energy Information Administration, available at https://www.eia.gov/tools/faqs/faq.php?id=99&t=3 (accessed 25 September 2016).

[3] Municipality use case, Echelon, available at http://www.echelon.com/assets/blt65a-02a03328f66e8/Municipality%20Use%20Case.pdf

[4] Caytiles, R.D. & Lee, S., A survey of recent power line communication technologies for smart micro grid. *International Journal of Software Engineering and Its Applications*, **9(12)**, pp. 251–258, 2015.
https://doi.org/10.14257/ijseia.2015.9.12.22

[5] Hrasnica, H., Haidine, A. & Lehnert, R., *Broadband Powerline Communications: Network Design*, Chichester, John Wiley & Sons, 2005.

[6] Vörös, B., *BPL technológia*, Master Thesis, Budapest University of Technology and Economics, 2012.

[7] Powerline Communications 'Emerge as Core Networking Technology, Electronic Design, available at http://www.electronicdesign.com/energy/power-linecommunications-emerge-core-networking-technology (accessed 25 September 2016).

[8] Yigit, M., Gungor, V.C., Tuna, G., Rangoussi, M. & Fadel, E., Power line communication technologies for smart grid applications: a review of advances and challenges. *Computer Networks*, **70**, pp. 366–383, 2014.
https://doi.org/10.1016/j.comnet.2014.06.005

[9] Sharma, D. K., Mishra, A. & Saxena, R., Analog & Digital Modulation Techniques: An overview. *International Journal of Computing Science and Communication Technologies*, **3(1)**, p. 2007, 2010.

[10] Amplitude Shift Keying & Frequency Shift Keying, available at http://www.ele.uri.edu/Courses/ele436/labs/ASKnFSK.pdf (accessed 25 September 2016)

[11] Understanding Modern Digital Modulation Techniques, Electronic Design, available at http://electronicdesign.com/communications/understanding-modern-digital-modulation-techniques (accessed 25 September 2016)

[12] What's the Difference Between Bit Rate and Baud Rate, Electronic Design, available at http://electronicdesign.com/communications/what-sdifference-between-bit-rate-and-baud-rate (accessed 25 September 2016)
[13] Third Generation Wireless Networks, Sky DSP, available at http://www.skydsp.com/publications/4thyrthesis/chapter1.htm
[14] Xiao, Y., Orthogonal frequency division multiplexing modulation and inter-carrier interference cancellation, 2003.
[15] Meel, J., Spread spectrum (SS), *Nayer Inst. Hogesch. Voor Wet. Kunst*, 1999.
[16] Carcelle, X., *Power Line Communications in Practice*, New York, Artech House, 2009.
[17] Ferreira, H.C., Grové, H.M., Hooijen, O. & Vinck, A.H., *Power Line Communication*. Wiley Online Library, 2010.
[18] Siong, A.K., *Design and Development of a Power Line Communication System*, The University of Newcastle, Australia, 2010.
[19] Berger, L.T., Schwager, A. & Escudero-Garzás, J.J., Power line communications for smart grid applications. *Journal of Electrical and Computer Engineering*, **2013**, p. 3, 2013.
https://doi.org/10.1155/2013/712376
[20] Papadopoulos, T.A., Kaloudas, C.G., Chrysochos, A.I. & Papagiannis, G.K., Application of narrowband power-line communication in medium-voltage smart distribution grids. *IEEE Transactions on Power Delivery*, **28(2)**, pp. 981–988, 2013.
https://doi.org/10.1109/tpwrd.2012.2230344
[21] Vukicevic, A., *Electromagnetic Compatibility of Power Line Communication Systems*, Switzerland, Lausanne, 2008.
[22] Bottyán, B., Király, T., Smart technológia alapú közvilágítás, *60. MEE Vándorgyűlés*, ID 5347, 2013.

TOWARDS THE DIGITAL PLATFORM AND SMART SERVICES FOR MANAGING SPACE TRAFFIC

PETR O. SKOBELEV & OLEG I. LAKHIN
Institute for Control of Complex Systems of the Russian Academy of Sciences,
Software Engineering Company "Smart Solutions", Russia.

ABSTRACT

The objective of the paper is to discuss the increasing complexity of modern space traffic in the near-Earth space and outline the new approach for solving the problem. The requirements and functionality of digital platform for traffic management are presented and examples of problem solving are given. The developed approach will create new opportunities for managing space traffic and resources of the mission control centers for a large number of spacecrafts. Possible approaches to description of spacecraft flights are given. Methods and tools for optimizing the use of ground control complexes to manage large-scale orbital groups have been discussed. Creation of the digital platform and eco-system of smart services for space traffic management will solve the most important problem of space traffic management to increase the effectiveness of the created satellites groups and to protect the spacecrafts from space waste and debris.

Keywords: Digital Platforms, Ground Stations, Management of Conflicts, Networks of Services, Planning and Control, Real Time, Satellites, Space Traffic, Spacecrafts.

1 INTRODUCTION

Currently, mankind is beginning to perceive outer space not only as an object of research and knowledge acquisition, but also as a new environment, the development of which brings commercial benefits [1]. It is already difficult to imagine life of modern human society without satellite-based navigation, mobile communication, digital maps compiled on the basis of space remote sensing of the Earth, etc.

Modern development of space systems and spacecrafts requires taking into account the following new factors:

- high speed of development of electronics and computer technology;
- development of additive technologies (3D printing) for design and production of spacecraft elements and systems;
- creation of terrestrial computing systems of high power and high reliability;
- development of reliable software used in space activities;
- extensive experience in flight control and operation of spacecrafts and spacecraft systems;
- significant progress in applied sciences sustaining space activities – ballistics, navigation, information processing, management of complex systems, etc.

In the nearest future we can expect the following important stages in development of space activities [2]:

- creation of a large number of cheap and relatively reliable spacecrafts, which have a small mass, small size and significant processing power;
- emergence of autonomous spacecrafts, in which most of the control tasks are carried out on board the spacecraft itself, and interaction with the ground control loop

DOI: 10.2495/DNE-V13-N2-187-198

is transferred into the area of determining long-term plans for operation of the spacecraft;
- emergence of multi-satellite systems, managed by the principle of "smart swarms";
- "explosive" increase in the number of simultaneously functioning spacecrafts in certain orbits or in a certain area of outer space;
- significant increase in the number of "space waste and debris" due to high intensity of space activities (broken-down spacecrafts, or those which have run out their operation term, unutilized remains of launch vehicles, fragments of collided spacecrafts, etc.);
- increase in the number of participants in space activities due to inclusion of new countries and new commercial companies, etc.

Thus, the boundless (as it seemed before) space, especially the near-Earth space, begins to become very "crowded", which can cause conflicts and reduce the effectiveness of space activities.

Currently the world community at the UN site starts initiative to develop basic rules for joint operations in outer space.

In this paper, we describe the proposed approach to developing space traffic control platform and eco-system of services, which is designed to allow adaptive and coordinated planning and control of spacecraft traffic in real time.

First of all, it is necessary to discuss the current state of the near-Earth space and those conditions in which space systems will operate in the near future.

2 INCREASING COMPLEXITY OF TRAFFIC MANAGEMENT IN THE NEAR-EARTH SPACE: "SPACE WASTE AND DEBRIS", EXPLOSIVE INCREASE IN THE COMPOSITION OF SATELLITE GROUPS, EMERGENCE OF NUCLEAR POWER

At present, many countries are forced to deploy their own control centers for the following objects in the near-Earth space:

- operating spacecrafts;
- upper stages of launch vehicles;
- spacecrafts that have stopped their active existence;
- old satellites, fragments of space vehicles and spent rocket stages, the fragments from their disintegration, erosion and collisions.

In addition, nowadays there is a large number of unmonitored or difficult-to-track objects in space:

- hundreds of thousands of space objects in the size range from 1 to 10 cm;
- hundreds of millions of space objects in the size range from 1 mm to 1 cm;
- billions of space objects less than 1 mm in size.

National centers, in their turn, assume availability of specialized micro-control centers for each spacecraft.

However, in case of any problem situation, strong coordination of actions between the teams of such centers is required. For example, the recent situation with a high probability of

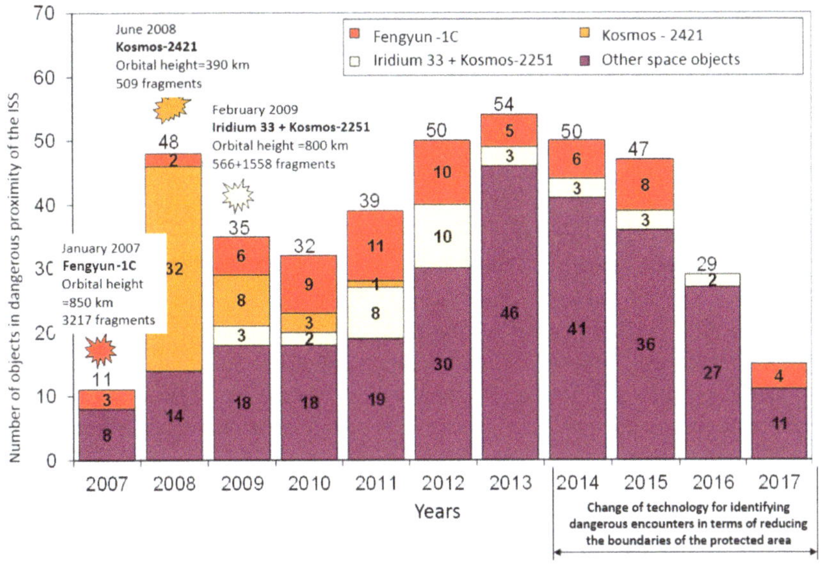

Figure 1: Number of dangerous encounters of the ISS with space debris

hazardous proximity of the spacecraft Kanopus-V with space debris: implementation of evasive maneuver plan required a whole range of activities and involved a whole number of additional specialists from related teams.

At the same time, the number of dangerous encounters with space debris continues to increase (Fig. 1) – a slight decrease in the right side of the chart is due only to the change of rules for registering such encounters.

Technogenic clogging of outer space is one of the most important problems of safe control of spacecraft flight.

If we consider distribution of large-scale space objects over the apogee altitudes and inclinations of their orbits, it is interesting to note that the most intensively clogged near-Earth space regions are in the range of altitudes from 800 km to 900 km and inclinations of 70° to 100°. Cases of destruction of spacecrafts as a result of collision with space debris, as well as ground impact of massive fragments have been recorded. The total mass of space debris in the near-Earth space is currently estimated at around 7,000 tons.

In Russia, automated warning system for dangerous situations in the near-Earth space (AWSDS NES / ASPOS OKP) has been created and is currently in operation. The system includes the main analytical center for monitoring dangerous situations in various areas of the near-Earth space, computation of parameters of solar and geomagnetic activity and analysis of problem situations about space objects [3].

The main tasks of the system are the following:

- continuous monitoring of the technogenic situation in the near-Earth space;
- identification, prediction and issuance of warnings about dangerous approaches of controlled spacecrafts with fragments of space debris, development of recommendations for conducting evasive maneuvers;
- forecasting the entry of uncontrolled risk objects into the atmosphere, determining the coordinates and time of their ground impact;

- analytical support of concerned parties with data on the technogenic situation;
- implementation of international obligations of the Russian Federation on problems related to space debris.

Systems like the Russian AWSDS NES are already one of the key elements in securing the safety of space flights in the near-Earth space and in the future the role of such systems will only increase.

Another serious challenge for developing control systems for future space systems is dramatic increase in the number of spacecrafts in the orbital space system. Thus, at present, orbital systems have several spacecrafts in their composition, and this is considered the norm. Orbital systems, consisting of several dozen spacecrafts, will be developed as a large space systems in nearest future including hundreds and thousands of spacecrafts. For example, a number of projects for providing global access to broadband Internet with the help of small spacecraft groups are now at various stages of implementation: OneWeb – 720 spacecrafts (+1972 spacecrafts in the long term), SpaceX – 4425 spacecrafts, Boeing – 2956 spacecrafts.

The quantitative transition from dozens to thousands of spacecrafts in space systems will undoubtedly bring the shift of paradigm in operational flight control and significantly influence both the methods and tools for operational management of the space system itself, as well as the types of coordination and interaction of various space systems [4].

Another potentially significant problem is the expansion of the use of nuclear energy in space and, consequently, collisions between space objects with nuclear sources on board and with space debris.

The beginning of the use of nuclear energy in space was marked on June 29, 1961 with the launch of the American spacecraft Transit 4A with the radioisotope thermoelectric generator SNAP-3B. Nuclear fuel, for example, plutonium-238, has long half-life (89 years) and relatively high energy release, which makes it very valuable for space energy.

One of the fundamental guidelines for the use of nuclear energy in space is that the organizations responsible for developing and launching spacecrafts with nuclear sources on board must make every effort to protect people and the biosphere from radiological hazards.

3 THE NETWORK APPROACH TO SPACECRAFT FLIGHT CONTROL: FROM CONFLICTS TO CONSENSUS

At present, space systems, as a rule, are de-facto created on the basis of a certain model of flight control. This model describes flight control as the process of changes in the parameters of the spacecraft state, aimed at achieving a given flight goal. Changes in the state parameters include a series of control operations aimed at changing the state of the spacecraft in accordance with the flight plan. The efficiency of such management requires to implement the set of operations in real and/or near-real time [5].

In the communication sessions with the spacecraft the flight control group continuously evaluates the state of the spacecraft. Based on the results of the assessment, a decision is made on the regular or abnormal condition of the spacecraft [6]. If the state is normal, control actions are taken to transfer the spacecraft to another state in accordance with the flight plan. However, if the state is abnormal, there starts the process of developing recommendations and such control actions that would allow the spacecraft to return to the planned state or, at least, to some state that meets the safety requirements of flight control. Such operational management implies successive execution of flight control stages: planning, implementation of the plan, monitoring and analysis of results, decision-making on control results.

Traditionally the control system is based on the feedback model, in which it is required to identify the state of control object, decide what actions to take, communicate actions with object and get feedback – to adjust control actions.

As a consequence of using this model in design of control solutions for space systems, in terms of flight control, the following features are required:

- methods and rules of flight control for each spacecraft;
- astro-ballistic models of spacecraft motion;
- specialized software for each spacecraft or space system;
- planning systems;
- integrated analysis of flight control processes and results.

However, with the beginning of mass exploitation of multi-satellite systems in conditions of the growing volume of space debris, the structures and processes arising during the flight control of various spacecraft groups are becoming increasingly connected with each other in a complex manner. Since these links are also able to dynamically change over time, specification of the flight control systems in the form of a classical control models with feedback ceases to satisfy practical requests.

It seems more preferable to have distributed network-based control processes of a satellite group as a whole [7], in which the network of earth-based users, ground and space stations and spacecrafts meet all needs (Fig. 2):

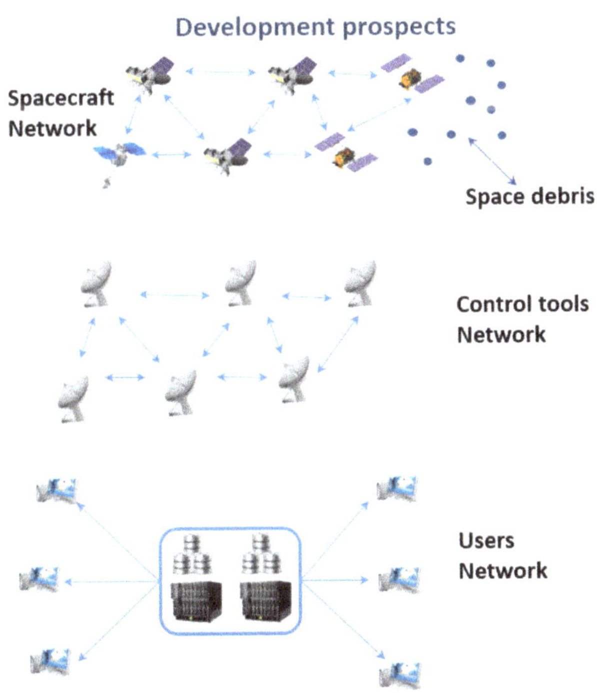

Figure 2: Network model of spacecraft control

1. Earth-based users have the opportunity to interact with each other, inform each other about their decisions, cooperate and coordinate with each other in achieving their goals, negotiate and compete with each other in the open information field.
2. The number, composition and use of ground and space flight control tools can be dynamically changed depending on the needs and capabilities of earth-based users, as well as on external causes (accidents, malfunctions, in-house maintenance operations, etc.)
3. The number and composition of the spacecraft group is dynamically changed due to entry of new spacecrafts and disposal of spacecrafts that have served their time or due to the consequences of emergency situations.

The use of a "networked" flight control model allows to formulate, at least at the conceptual level, the business requirements of a certain systems which will be a higher-order system (system of systems) in relation to the flight control system of a particular individual spacecraft. The objective of this system should be moderation of the spacecraft network, the network of space stations and management systems and assistance to earth-based users in developing effective interaction and negotiation processes [8].

The proposed network of systems will require the new adaptive resource management technology for organizing and coordinating space traffic [9].

Part of this technology could be blockchain for sharing resources "on demand" between different organizations (for example, ground stations).

4 DIGITAL PLATFORM AND ECO-SYSTEM OF SMART SERVICES FOR ADVANCED SPACE TRAFFIC MANAGEMENT

Advanced space traffic management (ASTM) involves digital platform for building an eco-system of smart services [10, 11] – to combine different models and methods for developing new generation of space systems.

The working paper submitted by the Russian Federation at the 59th session of the Committee on the Peaceful Uses of Outer Space (30.01.2017 – 02.10.2017), titled "Additional ideas on the set of goals to achieve Vienna consensus on safety in outer space and the need for thorough understanding on the modalities of consideration of complex issues related to space traffic management, and justification of the great expectations of early decisions in this field" may form a new legal basis for developing new space platform for cooperation, suggesting the following steps:

- to create and agree the concept of traffic management in space based on thorough study of all problematic issues and economic risks;
- step transition from ensuring safety of space operations at the level of documents of voluntary execution to the rules governing traffic in space;
- to create a single UN-based digital platform for monitoring and dissemination of information about objects and events in the near-Earth space;
- introduction of formal requirements for accuracy, completeness and timeliness of providing monitoring information about objects and events in the near-Earth space.

The main tasks of creating a digital platform for managing space traffic are the following:

- developing guidelines, standards and procedures relating to safety of space operations and avoiding collision of spacecraft with operating spacecrafts or elements of space debris;

- development and implementation of rules for provision and use of information on the situation in space;
- prevention of further generation of space debris and implementation of measures for safe cleaning of the near-Earth space from space debris;
- control of the state of the orbital group in order to identify dangerous flight situations and develop recommendations for their parrying in operational mode;
- provision of continuous updating of the space-time model of the near-Earth space;
- development of recommendations on planning of orbital operations in order to improve safety of space systems traffic and movement of individual spacecrafts in the current and forecested space environment;
- timely, complete, authentic and accurate communication of information about objects and events in the near-Earth space to consumers;
- carrying out analysis of processes of managing spacecrafts as a part of space traffic control technology;
- participation in planning and preparation for the launch of new spacecrafts and space systems in order to assess the risks for the expected period of operation, justification of acceptable orbits, correction of time for the launch of carrier rockets and boosters.

Based on these tasks, the following key “smart services” can be developed (Fig. 3):

- smart services for ballistic models;
- smart services for models and methods of data analysis;
- smart services for planning at all levels.

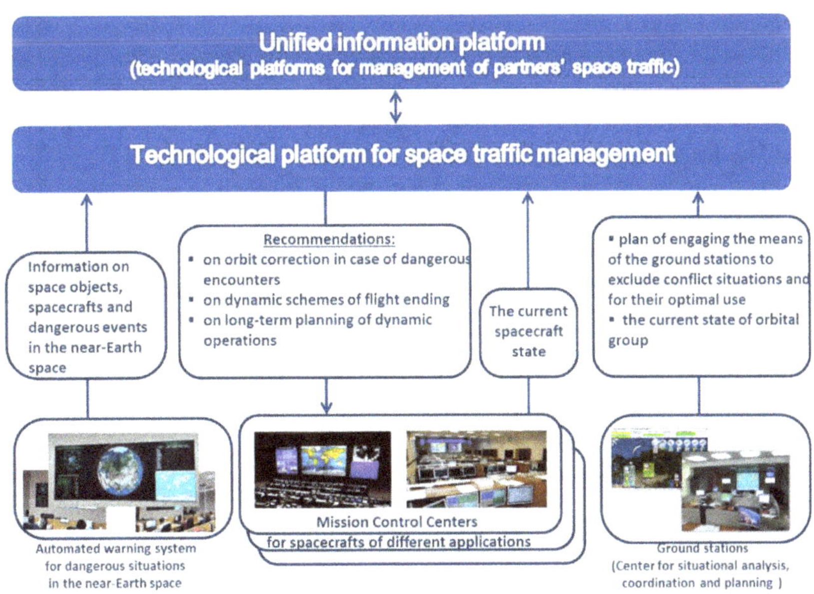

Figure 3: Structure of the digital platform of space traffic management

The features of any "smart service" should be the following:

- Ontology-driven Knowledge base – which should contain a semantic network of concepts and relations of the domain area [12];
- Autonomy – the ability to continuously respond to events, make plans and achieve their fulfilment.

The platform will be based on mathematical models and methods of ballistics, which should provide the following possibilities:

- creation of a ballistics-navigation support system for flight control for a large number of multi-type spacecrafts controlled from a single mission control center;
- methods of project-ballistic analysis and expertise of orbital construction of space remote sensing systems, communication and retransmission;
- development of software for selection and calculation of parameters of orbits for disposal of spacecrafts;
- improving methods for calculating dangerous encounters and parameters of evasive maneuvers for spacecrafts;
- methods of ballistic design of space complexes for orbiting maintenance of spacecrafts in the near-Earth orbits;
- improvement of methods for calculating ballistic schemes for removal of large-sized space debris;
- comprehensive methodology for predicting technogenic pollution of an area of 1100 (up to 2000) km with objects larger than 10 cm;
- development of a methodological approach to assessment of safety of payload generation scheme;
- improvement of methods and technologies for traffic control and prediction of dangerous encounters;
- development of methods and technologies improving the accuracy of forecasting the time and impact point of unmanaged space objects;
- study of the evolution of "clouds" of space debris, formed as a result of destruction of rocket and space equipment objects in the low-earth orbit;
- improving interaction between organizations engaged in safety analysis and preparing and launching space-mission vehicles;
- creation of space observing devices in the visible and infrared ranges for fragments of space debris in the low-earth orbit;
- measures for prevention of the risk of collisions of functioning spacecrafts with formed fragments.

The development of the platform will make possible to significantly improve the quality and efficiency of space traffic management in the context of constantly growing technogenic pollution of the near-Earth space, namely:

- maintenance of space vehicles of the Russian orbital group, as well as identified potentially dangerous space objects and risk objects;
- control of implementation of the stages of spacecraft launch into target orbits, measures to remove the spent Russian spacecrafts, the upper stages of launch ve-

hicles and boosters into the "dead" zones and into orbits with a limited life, assess parameters of the state of accompanied spacecrafts (including in non-emergency situations);
- assessment, identification, forecast, analysis, and ballistic support of dangerous situations, including destruction of space objects in the near-Earth space, dangerous encounters of spacecrafts with potentially dangerous space objects, forecasting descent from orbits and ground impact of risk objects identifying time and possible impact area.

The next key element of the space traffic management is the data analysis services – aimed to improve technologies for analyzing flights parameters of spacecrafts and their control systems, forecast their functioning on the basis of accumulated knowledge base of flight data, as well as monitor the state of elements of the ground stations.

Figure 4 shows the structure and integration of these services.

The main functions of the data analysis services are the following:

- collection, processing and storage of information on flight data of individual spacecrafts and spacecraft systems;
- forecast of operation of spacecrafts and space systems based on the accumulated knowledge base of flight data;
- monitoring the state of the elements of the ground stations.

The proposed approach provide solution for solving the problems of telemetric and information support for flight control of manned and automatic spacecrafts and space systems for the benefit of creating the unified space traffic management technologyfor a large number of spacecrafts.

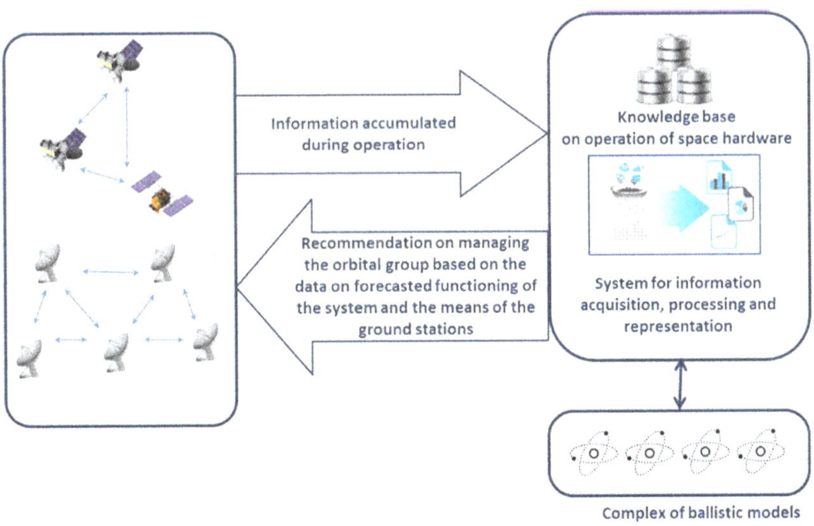

Figure 4: The structure and integration of data analysis services

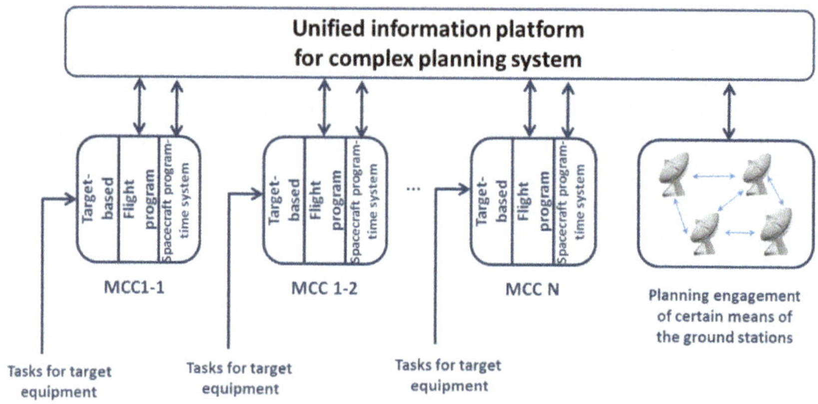

Figure 5: Planning services based on the digital platform

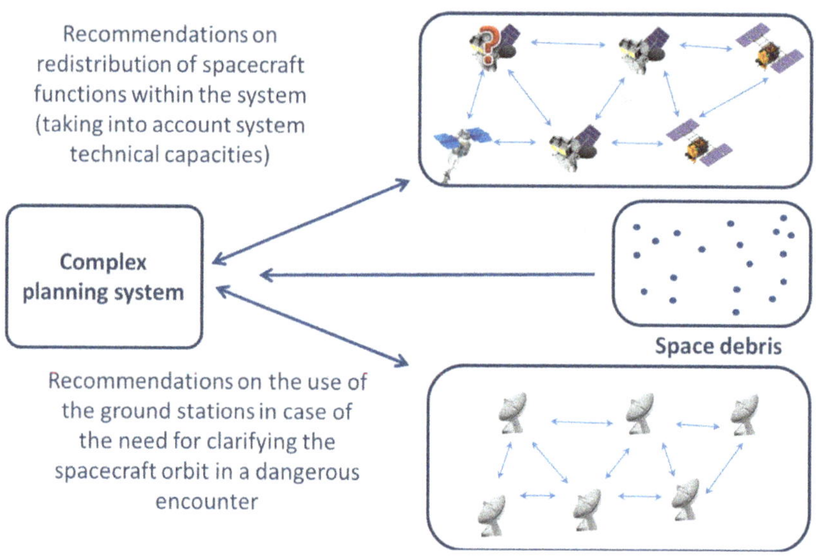

Figure 6: Scheme of task redistribution in case of the event of an abnormal situation

In order to create services for strategic and operational management of large-scale orbital groups of spacecrafts, it is necessary to use models and methods for allocating resources of ground-based and satellite spacecraft control tools, as well as mutual maneuvering of spacecrafts within the existing space technology capabilities (Fig. 5).

With the use of developed platform it becomes possible to provide modernization of the Center for situational analysis, coordination and planning to manage a large number of different types of spacecrafts, as well as to support interaction between spacecrafts and ground stations for scientific, social and economic applications.

At the same time, an end-to-end network of spacecrafts and stations can be provided for managing abnormal situations (Fig. 6).

In addition, within the platform, there should be implemented services for target planning for groups of various types of spacecrafts, services for mamaging of the spacecraft flight program, services for space-time control of spacecrafts, services for planning the use of individual elements of the ground stations and services for forming tasks for the target equipment. The platform should support organization of tasks redistribution in the spacecraft system in cases of unpredictable events of an abnormal situation and the need for unscheduled evasion from space debris. At the same time, additional services should be implemented for issuing recommendations on redistribution of spacecraft functions within the system (taking into account the technical capabilities of the system) and for issuing recommendations on the use of the ground stations for clarifying the spacecraft orbit in case of a dangerous encounter.

Another important issue in developments of the space traffic management is legal and regulatory framework. Since the space traffic management technology affects aspects of international cooperation, it is necessary to take into account not only the Russian GOST (State All-Union standard), but also international ISO, CCSDS standards, and documents of the Inter-Agency Space Debris Coordination Committee (IADC).

CONCLUSION

The objective of the developed approach is to create new opportunities for managing space traffic and resources of the mission control centers for a large number of spacecrafts with different owners.

The proposed space traffic management platform and eco-system of smart services will provide a comprehensive solution of the following tasks:

- continuous round-the-clock control of automatic spacecrafts in various orbits;
- cataloguing information on space debris in the near-Earth space;
- rapid identification of dangerous encounters;
- determination of the time and area of ground impact for space objects;
- coordinating the work planning for the ground stations and for spacecrafts of scientific, social and economic applications in order to exclude conflict situations and their optimal use;
- analysis of the use of the ground automated control complex for spacecrafts of scientific, social and economic applications;
- analysis of the status of the orbital group of spacecrafts of scientific, social and economic applications;
- provision of round-the-clock continuous control of the Russian Segment of the ISS RS;
- providing control of Transport Cargo Vehicles of the "Progress MS" and "Soyuz MS" types;
- ensuring the conduct of scientific experiments.

Currently, methods and tools for developing digital platform for control spacecrafts and and use of ground stations are under development.

The main requirement is to manage large-scale orbital groups of up to 40 spacecrafts at the existing ground stations including small, ultra-small and nano spacecrafts that will be formed in a self-organizing "swarm" or "cloud" –to make the Roskosmos orbital group more smart, open and flexible, scalable, efficient and reliable and also protect the spacecrafts from space waste and debris.

ACKNOWLEDGEMENTS

The paper has been prepared based on the materials of scientific research within the subsidized state task of the Institute for Control of Complex Systems of the Russian Academy of Sciences for research and development on the topic: "Development of models, methods and tools for supporting and developing mechanisms of multi-level self-organization for improving task and resource management in multi-discipline projects of manufacturing complex technical products".

REFERENCES

[1] Lobachev, V.I. & Pochukaev, V.N., "Methods of testing rocket and space equipment", Mechanical Engineering, 1995. Section "Means of flight control of scientific and manned space vehicles".

[2] Matyushin, M.M., Sokolov, N.L. & Zakharov, P.A., Features of control of large-scale groups of space vehicles. *Problems of Control and Modeling in Complex Systems*, 2016.

[3] Stepanov, I.B. & Pavlova, S.A., Events in near-Earth space // TsNIIMash, Information and Analytical Support of ASPOS OKP (automated warning system for dangerous situations in the near-Earth space). *The Main Information and Analytical Center of ASPOS OKP*, **9(76)**, 48 p.

[4] Udaloy, V.A. & Sokolov, N.L., On the possibility of mathematical formalization of the functioning processes of the Mission Control Center as a large control system. *Modern Science-Intensive Technologies*, 2006.

[5] Matyushin, M.M., Sokolov, N.L. & Ovechko, V.M., Methodological approaches to solving problems of managing large-scale groups. *Engineering Journal: Science and Innovation*, **51**, 2016.

[6] Matyushin, M.M., Potockiy, S.I., Skobelev, P.O., Potapov, V.I. & Lakhin, O.I., Automated decision-support system for emergency situations. *Software Products and Systems*, **(3)**, pp. 62–69, 2013.

[7] Matyushin, M.M., Sokolov, N.L., Ovechko, V.M. & Polivnikov, V.M., Designing rational schemes for distribution of control means during operation of large-scale orbital groups of space vehicles. *Flight. All-Russian Scientific and Technical Journal*, 2016.

[8] Rzevski, G. & Skobelev, P., *Managing Complexity*. Southampton, Boston, WIT Press, 2014.

[9] Skobelev, P.O., Lakhin, O.I., Polnikov, A.S. & Simonova, E.V., Approach to the solution of aerospace product lifecycle management problem based on network-centric principles. *Proceedings of the 7th International Conference on Industrial Applications of Holonic and Multi-Agent Systems* (HoloMAS'2015), eds. V. Marik, J.I. et al., , Valencia, Spain. – HoloMas 2015, pp. 169–178, September 2–4, 2015.

[10] Digital ecosystem, avialbale at https://en.wikipedia.org/wiki/Digital_ecosystem (accessed 18 March, 2018).

[11] Skobelev, P., Multi-agent systems for real time adaptive resource management. *Industrial Agents: Emerging Applications of Software Agents .in Industry*, ed. Paulo Leitão & Stamatis Karnouskos. Elsevier, pp. 207–230, 2015.

[12] Skobelev, P., Activity ontology for situational management of enterprises in real-time. *Ontology of Designing*, **1(3)**, pp. 6–38, 2012.

LEAN AND COMPLEX SYSTEMS: A CASE STUDY OF MATERIALS HANDLING AT AN ON-LAND WAREHOUSE FACILITY SUPPORTING SUBSEA GAS OPERATIONS

PER ENGELSETH & DANIELA GUNDERSEN
Department of Logistics, Molde University College, Norway.

ABSTRACT

Commonly warehouse management is subject to deterministic planning procedures within the organizational scope of supply chain management. Lately lean principles have become more widespread and also adapted to develop materials handling within warehouses. Several facets of analysis are considered in this case study on outsourced warehousing in the Norwegian petroleum industry. This study reveals empirically how outsourced warehousing as service provision involves an organization problem with waste found both at the service management and at the process levels. Analysis points to how lean principles of process improvement may be combined with complex systems thinking. The case reveals examples of how warehouse management and its operations are full of complexities leading to waste. While lean usually is used hands on to manage these wastes through manual incremental development, another organisation of continuous innovation is suggested here. Lean is suggested be developed as organizational context of complex systems information warehouse management system software. Following lean principles evokes many types of process waste at the studied warehouse. Instead on following the manual stepwise and continuous process improvement approach of lean in all areas of organization, a more radical change to complex systems evident in information system design to support warehousing processes is proposed with lean as its organizational context.

Keywords: case study, complex systems, lean, material handling, Norwegian petroleum industry, services industry, Warehouse management.

1 LEAN AND ITS COMPLEXITY

This paper juxtaposes the concepts of "Lean" and "Complex Systems" in the context of supply chain management (SCM) as principles of production. To juxtapose concepts means a search for contrasting effects through interrelating these words in a text. Meanings may thus be developed as a conceptual exercise. These are two concepts, "Lean" and "Complex Systems" when used within SCM are thus considered associated in with different approaches on how to manage flows of people, goods and information. Herein lies the academic value of this exercise.

The key to efficient SCM is associated with, following the seminal works of Forrester [1], the integration of differentiated and sequentially interdependent actors in the supply chain to avoid production inefficiencies through e.g. bullwhip-like effects. Information distortion is a key challenge in SCM and collaboration is deemed the main remedy to this to secure logistics quality in the form of handing down to the end-user in the supply chain service meeting customer expectations regarding the time, place and form characteristics [2]. In relation to aims of creating a true "value chain" out of just any other "supply chain", the notion customer perceptions need to be significantly elevated. These notions are all-important when considering Lean and Complex Systems in a SCM context. This sets forth our understanding of what constitutes "Lean" and how lean is associated with "complex systems" thinking. Juxtaposing these two concepts implies also respecting their integrity as different, not forcing them to merge into some new concept. This implies accordingly that this paper seeks ways to use

DOI: 10.2495/DNE-V13-N2-199-207

Figure 1: The Nyhamna facility

these concepts to improve materials handling at a same time, but keep the concepts unchanged. The following text provides introductions to "lean thinking" and "complex systems" and "services" as basis for discussion about the empirical findings in section six.

The case concerns warehousing management at the Nyhamna facilty in North-western Norway. The well stream from Ormen Lange, Norway's second largest gas field, is processed at the land facility in Nyhamna since 2007, owned and operated by Shell, where the gas spends just about ten minutes before it is exported 1.200 km. through one of the world's longest subsea pipelines, Langeled, to Easington in England. Shell is now expanding this facility involving suppliers of materials and components from all over the world. Among these suppliers is our focal firm we call Aurora. The analytical framework ids developed through first introducing lean thinking, followed by complex systems and warehousing services.

2 LEAN THINKING

Lean management principles evolved from Taylor's scientific management through Ford's principles of mass production, continuing through Ohno's just-in-time manufacturing system found in the Toyota Production System [3], into an increasingly knowledge-based humanistic system for process development setting people in focus as basis for industrial manufacturing process improvement [4, 5]. The term "lean" is itself a relatively recent invention. It was first coined by John Krafcik in a fall 1988 article, Triumph of the Lean Production System, published in the Sloan Management Review, based on his master's thesis at the MIT Sloan School of Management. The term "lean" amplifies the core objective of the Toyota Production System [3] that is associated with elimination of waste, or *muda* as it is called in Japanese. One of the key routes to process development through lean thinking is to understand the present based on the past; thereby improve future production. Extensive, and potentially continuous modelling of existent production processes to reduce the mainly time-associated muda lays ground for incremental change. Teamwork is an important enabler of such development, and the role of management is to enable this change; the notion of "servant leadership". Lean is accordingly a predominately manual approach to process change and values interaction set in norms to develop production to economise production.

3 COMPLEX SYSTEMS

Complexity is defined as "…a property of an open system that consists of a large number of diverse, partially autonomous, richly interconnected components, often called agents, has no centralized control and whose behaviour emerges from the intricate interaction of agents and is therefore uncertain without being random" [6; p. 5]. Rezevski and Skobelev [6] continue by pointing out that the key features of complexity are openness, diversity, partial autonomy and interconnectedness of agents, lack of centralized control and emergence. This interlinking of components entails interdependence. In a complex system management must consider how to manage processes where components are continually in change regarding not only logistics; the time, place and form features, but also how they interact as well as how this interaction is perceived in a value chain. In a complex system, resource objects are not only transformed through production, but their evaluation may change as time unfolds. Complexity entails at core emergence. This is the focus of analysis in complex systems thinking. "Generative emergence" entails change found in organizations and therefore inclined to be managed [7]. A key attribute of process development following complex systems logic is supporting connectivity through *information systems* designed explicitly to manage generic emergence in processes. There is an inducement when developing supply following complex systems logic towards automating this interaction by designing software that enables navigation as opposed to deterministic planning.

4 WAREHOUSEING SERVICES

Service management was initially developed conceptually through the services marketing literature [8, 9]. Particularities of services supply include that services are intangible, heterogeneous, perishable, and inseparable regarding supplier-customer interaction [10]. Grönroos [8] highlights particularities of service production including that services at some extent services are impalpable. This means that they consist of a sequence of activities rather than objects, that services can be generated and consumed contemporaneously and that the customer often takes part in the service production process. Service processes are therefore inherently complex in nature, possibly more complex than manufacturing processes which represent the empirical foundation for the conceptual development of Lean principles.

In its generic form, services are commonly classified as intangible, heterogenic, inseparable, and perishable. This static classification provides, according to Spring and Araujo [11], limited analytical value since it does not consider interactions involved in services production. This implies that producing services is organized differently than in manufacturing [12, 13]. People are fundamental to achieving quality in services [9, 10]. Sampson and Froehle [14] state that: "With service processes, the customer provides significant inputs into the production process". This underpins the importance of interaction and that value in service is inherently to some degree co-created. A service cannot start before the customer informs the supplier with its requirements. This requirement may then also be negotiated.

According to Sampson and Froehle [14], three types of customer inputs can be found in services: (1) the customer person; (2) physical resources such as customer belongings, tools and other tangible objects; and (3) information. These resources are pooled and in combination produce a service. Another distinct feature of service supply is that these supply chains can be characterized having a non-linear "hub and spoke" network-like structure. Sampson and Froehle [14] state that seldom more than three tiers of firms are found in a service supply chains; it is in principle a triadic form of collaboration.

A simple way to analyse services from a strategic viewpoint is to apply interdependency theory [15]. According to Thompson's [15] framework, all forms of interdependencies, sequential, pooled and reciprocal, are accountable in all types of supply; it is their importance that varies based on an evaluation if value chain power structure. Complete industries often show similarities regarding their power structure, making it sensible to consider how industrial type particularities impact on interdependencies. Services are characterized by either predominately pooled or reciprocal interdependencies.

Following Thompson [15], Stabell and Fjeldstad [16] propose a framework where services may be classified as value shops where reciprocal interdependencies are dominant, or value networks where pooled interdependencies are dominant. This implies that within services, two main types of interdependencies may be found as power structure. When developing service processes muda (waste) is not predominately found in how operations are timed sequentially (as in the lean principle of "tact time") but how resources are pooled and how actors interact. In services muda is associated with either mainly pooled or reciprocal interdependency. Intensive interaction to mutually adjust production is found in reciprocally interdependent relationships. When pooled interdependencies are predominant, connectivity achieved through production resource standardization (including people to support resource mediation) using mediating technology needs to be developed.

Waste in services concern e.g. random arrivals, inconsistent specification, and varying input quality that influence service processes as capacity, demand management and quality management [14]. Bicheno and Holweg [17] discuss, from a lean perspective, various forms of muda found in services. These are: (1) delay, (2) duplication, (3) unnecessary movement, (4) unclear communication, (5) incorrect inventory, (6) poor customer service, and (7) transaction and production errors. Following lean thinking, these problems should be solved through incremental change that is founded on worker empowerment, so that they notice and are motivated to drive change. From complex systems thinking, change should involve fundamentally changing the system structure, especially its information system, to create a supporting context helping agents navigate through rather than plan themselves away from uncertainty and experienced challenging events emerging as warehousing production unfolds following the timeline as descriptive mode.

5 METHOD

Without a research focus it is easy to become overwhelmed by the volume of data in qualitative research [18]. In this case of warehouse management, the studied processes involve details regarding how and why people acts as they do in a warehouse management context. Only an abstract of these findings are presented in the following narrative.

Following Miles and Huberman [19] and Yin [20] the case study research strategy was used because it provided grounds to explore warehouse processes in their real context. Taylor and Fearne [21], Fernie and Thorpe [22] and Holweg and Pil [23] state that case studies is appropriate when describing actors, structure and agency relations taking place through social interaction in further detail. Case studies (1) make it possible to answer "how" and "why" research questions, (2) researchers cannot manipulate the behaviour involved during the research process and (3) researchers can seek a picture of the context the phenomenon is embedded in.

This is a single case study. This entails that process details witting the case are reported [24], as opposed to comparison between cases. According to Meredith [25] direct observation is more important than second or third observation. This also implies retrospective recollection of events and processes, a limitation and potential bias since things may be perceived wrongly and recollections may be distorted over time. Any researcher who participates in an organization over an extended period of time will find it is difficult not to participate actively in the organization [26]. The main risk of participant observation as a method for collecting data is the danger of "going native" and loses the perspective necessary for a researcher.

One of the researchers, working in the company in question, had therefore ample access to documentation, as well as potential to discuss findings with co-employees, workers as well as management. Cross reference to secure credibility was done through informal interviews with different employees here they describe the situation before September 2016, when they had only 5–6 people working there.

6 CASE TEXT

The expansion project is large and Aurora has a central role in the project. More than 160 workers are present on site, divided on different departments, disciplines and locations. Thus is an organization problem. The amount of work is large, the products used are many, the activities are complex. Due to the size, the need for communication and interaction is high and coordination is complicated. Overlapping project activities between engineering, production and warehouse department: Some of the tasks are not clearly defined from the beginning. This creates extra work and waste of time and resources. Maturity of design/technology, meaning some of the products doesn't fit the gas pipes. This demands more coordination, since a number of adjustments are necessary.

The case narrative focuses on describing operational weaknesses in warehousing processes. The warehouse processes are shown in Fig. 2:

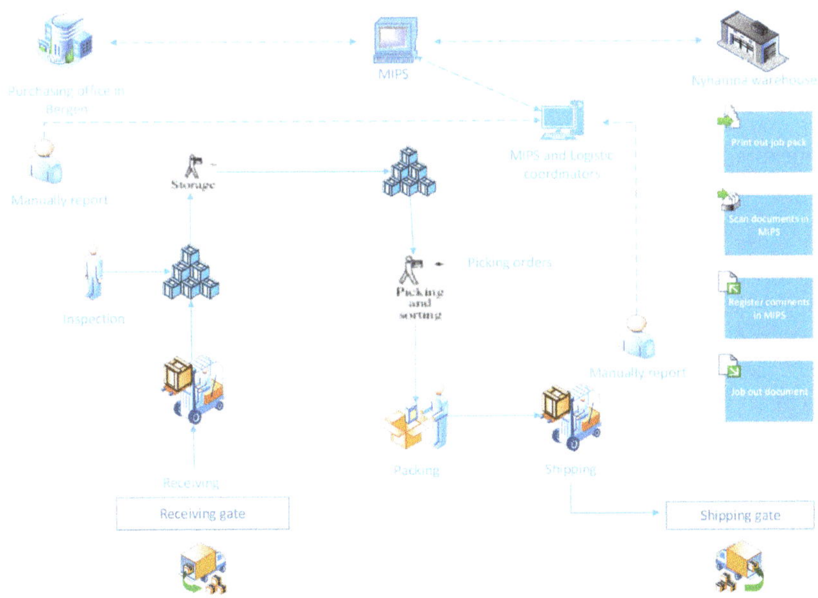

Figure 2: Aurora warehousing process

Prior making the adjustments, the production staff has to interact with supplier personnel. After the products are modified, the supplier contacts the warehouse. All three departments are in close dialog. Such interactions tend to be very time consuming. The situation can generate delays during the project execution. Collaboration between project partners contributes to the improvement of the services that the warehouse delivers. The communication with the contractor's transport department improved after the warehouse personnel increased and the tasks are better distributed. The same thing happened with other suppliers that handled the transport of new purchased goods. Operational collaboration is a factor that helps coordination and better flow of the project.

Process weaknesses associated with material handling at the Aurora Nyhamna facility are many. After confirming the delivery, the system is not good enough to follow up what happen with the goods. The items are packed, but not delivered. The goods are stored at warehouse and the only way details are registered by manually using the comments in the warehouse management system (MIPS). The workers need to fill out a form whenever they pick up materials. The information on this form is, however, incomplete missing details such as item number, in-warehouse location and volume.

Poor communication is found between foremen, team leaders, General Foremen (GF) and the production department. As an example, the GF asks warehouse to send goods, but the foremen is not informed about the ownership of that job; who is responsible for its operation. Another situation occurs when both Team Leader and GF send same request to warehouse, by making double registration in GF list. Foremen don't use their time to check status on goods. Instead, they prefer to send many e-mails to the warehouse asking for information they could find on their own. Sometimes the foremen forget that they signed and received goods, asking for it again. The attachments in MIPS with signature help the warehouse personnel to prove that goods were sent and received.

Poor handover notes abide from one work-shift to another. Due to the stress and pressure they are faced to, the departments are pointing fingers at each other trying to avoid taking responsibilities for their acts. The production workers don't use enough time to look for goods on site. They prefer that warehouse uses the resources instead.

Several factors that affect performance at the warehouse include integration of engineering, production and warehouse departments. These are located at the same site. Such a level of integration demonstrates an important factor in facilitating coordination. They have direct communication, but sometimes this is challenging because there are too many people involved (workers worked in shifts, changing often without perfect handover notes).

7 CONCLUSION

Based on this description some propositions for merging lean thinking with complex systems in the context of warehousing management emerge. Materials handling involves a set of what intuitively should be considered relatively simple processes involving receiving goods, handling them at the warehouse for cross-docking or storage, picking and dispatching them. Lean challenges practitioners to *continuously* never forget to push for improvement. It is concerned with understanding the root causes of muda.

Lean is the people-focused tough highly manual path to economizing e.g. warehousing management. Applying lean thinking concerns taking small steps at a time meaning following a line of incremental deployment. Within lean thinking error detection involves people perceiving and in personal modes reporting and through teams working on this. Improvement rests on the shoulder of the people in a company, both labour and management, including how well these are culturally integrated, share a common organizational discourse.

Neither the studied company, Aurora nor any of their co-creators in the value chain apply complex systems thinking as such. They do follow lean principles to some degree, enough to understand that the status quo of the described warehousing is not satisfactory. Complex systems can provide an alternative structure to improving the process of warehousing as everyday operations. Mainly, this would imply to start with a form of radical change. This is different from lean thinking and its focus on incremental change. Lean is not a radical process improvement strategy. It takes the safe path to organizational development. If one step ahead is wrong, this step can easily be corrected. Investing in complex systems software would, however, be a disruptive radical innovation to the studied warehouse management systems.

Implementing complex systems thinking, however, would demand investing in an information system and production resources where focus would be on system flexibility and interconnectivity. Lean, with its focus on incremental change, also has no prominence for deterministic long-term supply planning. Complexity, defies linear logic as it brings with it self-organization and feedback loops, wherein the effect is its own cause [27]. In a manner Lean thinking therefore implies a manual variation of complex system due to its line of incremental continuous improvement. This complexity is, however, not immediate since it demands human intervention. This may take some time and demands some form of human effort. This may itself be a form of muda, waste associated with process improvement. It is at this organizational layer, surrounding use of complex systems software in warehouses, that lean principles are suggested be applied.

In sum, complementarity is accordingly suggested applied to mix these methodologies for process improvement. Lean is suggested as organizational back-up for more automated warehouse processes applying complex systems methodology. Further research may be dedicated to reporting on such cases of process improvement within a wide range of logistics applications.

REFERENCES

[1] Forrester, J., *Industrial Dynamics*. Waltham, MA: Pegasus Communications, 1961.

[2] Alderson, W., *Dynamic Marketing Behavior. A Functionalist Theory of Marketing*, Homewood IL, Richard D. Irwin, 1965.

[3] Ohno, T., *Toyota Production System. Beyond Large Scale Production*, Boca Raton FL, CRC Press, 1988.

[4] Womack, J.P., Jones, D.T. & Roos, D., *The Machine That Changed the World: The Story of Lean Production*, New York, Rawson Associates, 1990.

[5] Holweg, M., The genealogy of lean production. *Journal of Operations Management*, **25**(2), pp. 420–437, 2007. https://doi.org/10.1016/j.jom.2006.04.001

[6] Rzevski, G. & Skobelev, P., *Managing Complexity*, Southampton UK, WIT Press, 2014.

[7] Lichtenstein, B.B., *Generative Emergence*, Oxford, UK, Oxford University Press, 2014.

[8] Grönroos, C., *Service Management and Marketing: Managing the Moments of Truth in Service Competition*, Lexington MA, Lexington Books, 1990.

[9] Normann, R., *Reframing Business: When the Map Changes the Landscape*, West Sussex UK, Johan Wiley & Sons, 2001.

[10] Ellram, L, Tate, W. & Billington, C., Services supply management: the next frontier for improved organizational performance. *California Management Review*, **49(4),** pp. 44–66, 2007.
https://doi.org/10.2307/41166405
[11] Spring, M. & Araujo, L., Service, services and products: rethinking operations strategy. *International Journal of Operations & Production Management*, **29(5)**, pp. 444–467, 2009.
https://doi.org/10.1108/01443570910953586
[12] Chase, R. & Garvin, D., The service factory. *Harvard Business Review*, **6(4)**, pp. 61–69, 1989.
[13] Oliva, R. & Kallenberg, R., Managing the transition from products to services. *International Journal of Service Industry Management*, **14(2)**, pp. 160–172, 2003.
https://doi.org/10.1108/09564230310474138
[14] Sampson, S.E. & Froehle, C.M., Foundations and implications of a proposed unified services theory. *Production and Operations Management*, **15(2)**, pp. 329–343, 2006.
https://doi.org/10.1111/j.1937-5956.2006.tb00248.x
[15] Thompson, J.D., *Organizations in Action*, New York, NY: McGraw Hill, 1967.
[16] Stabell, C. B. & Fjeldstad, Ø. D., Configuring value for competitive advantage: on chains, shops, and networks. *Strategic Management Journal*, **19(5)**, pp. 413–437, 1998.
https://doi.org/10.1002/(sici)1097-0266(199805)19:5<413::aid-smj946>3.3.co;2-3
[17] Bicheno, J. & Holweg, M., *The Lean Toolbox: The Essential Guide to Lean Transformation*, Buckingham UK, PICSIE Books, 2009.
[18] Eisenhardt, K.M., Building theories from case study research. *The Academy of Management Review*, **14(4)**, pp. 532–550, 1989.
https://doi.org/10.5465/amr.1989.4308385
[19] Miles, M.B. & Huberman, A.N., *Qualitative Data Analysis*, Thousand Oaks, CA: Sage, 1994.
[20] Yin, R.K., *Case Studies Research: Design and Methods*, Thousand Oaks, CA: Sage, 2009
[21] Taylor, D. & Fearne, A., Towards a framework for improvement in the management of agri-food SCs. *Supply Chain Management: An International Journal*, **11(5)**, pp. 379–384, 2006.
https://doi.org/10.1108/13598540610682381
[22] Fernie, S. & Thorpe, A., Exploring change in construction: supply chain management. *Engineering, Construction and Architectural Management*, **14(4)**, pp. 319–333, 2007.
https://doi.org/10.1108/09699980710760649
[23] Holweg, M. & Pil, F.K., Theoretical perspectives on the coordination of supply chains. *Journal of Operations Management*, **26(3)**, pp. 389–406, 2008.
https://doi.org/10.1016/j.jom.2007.08.003
[24] Voss, C., Tsikriktsis, N. & Frohlich, M., Case research in operations management. *International Journal of Operations and Production Management*, **22(2)**, pp. 195–219, 2002.
https://doi.org/10.1108/01443570210414329
[25] Meredith, J., Building operations management theory through case and field research. *Journal of Operations Management*, **16(4)**, pp. 441–454, 1998.
https://doi.org/10.1016/s0272-6963(98)00023-0

[26] Åhlström, P. & Karlsson, C., *Longitudinal Field Studies, in Researching Operations Management*. London: Routledge, 2009.

[27] Jay, J.J., *An Introduction to Systems Thinking, The Ecosystem Approach*, eds. D. Waltner-Toews, J.J. Kat and N.-M. Lister, Columbia University Press, pp. 3–11, 2008.

ENSURING RESILIENCE AND AGILITY OF COMPLEX ORGANIZATIONAL-TECHNICAL SYSTEMS

L.D. GITELMAN, T.B. GAVRILOVA, M.V. KOZHEVNIKOV & E.M. STARIKOV
Department of Energy and Industrial Management Systems, Ural Federal University, Russia.

ABSTRACT

Modern organizational and technical systems have been developing in an environment that is marked by capriciousness, uncertainty, risk, variability, and evolution (CURVE factors). As organizational-technical systems grow bigger, their internal complexity increases, too, both structurally and dynamically. The article substantiates the appropriateness of employing the principles of systems engineering for managing such systems.

The authors analyzed various theoretical concepts of and practice-based approaches to the development of systems engineering in the context of ensuring the resilience and agility of complex organizational-technical systems. Using the case of power engineering and hi-tech industries, the authors show that for organizations that operate critical infrastructure facilities it is essential to make sure that the system stays functional in adverse conditions and is able to recover quickly after a failure. It is demonstrated that for addressing the above task it is critical to use instruments that nurture interdisciplinary competences in individual professionals and in teams that manage the development of complex systems and implement major innovation projects.

As part of the study, the authors also look at the possibility of using the principles of resilient systems design and the fundamental principles for agile systems engineering when managing critical infrastructure facilities.

Keywords: agility, critical infrastructure, CURVE factors, organizational and technical system, resilience, systems engineering.

1 INTRODUCTION

The authors define an organizational-technical system (OTS) as a network of groups of professionals from various domains that interact between each other and, despite possibly being part of different business structures, work for the same goal that is usually associated with the development of a system (business), exploration of new technological areas and markets, and adoption of breakthrough technologies.

In literature, the term "organizational and technical system" is often used a synonym of "socio-technical system" that was first introduced by E.L. Trist in the early 1960s [1]. However, the two definitions have a number of differences that are significant in the context of this study. For example, [2] notes that socio-technical systems focus on the interests and needs of the person and their relationships with technical aids (in terms of ergonomics), whereas in OTSs the emphasis is placed on managing the operation and development of such systems (technical guidelines, legal framework, organizational structures, business processes).

A socio-technical system assigns the same priority for *current* production performance indicators and employee work performance, while in OTSs the emphasis is shifted towards sustainable development of the system on the basis of projections of *its future* internal and external parameters (organizational links, development management) [3]. Another difference is that the organizational and technical subsystems within an OTS are intermingles, whereas in socio-technical systems each of the constituent subsystems has rather clear borderlines despite their interrelation [4].

Apart from the attributes that are typical of complex systems in general, a complex organizational and technical system has the following features [5–8]:

DOI: 10.2495/DNE-V13-N2-208-220

1. an active role of man-machine complexes that develop in a technological breakthrough mode and make it possible to launch production and provide services that use smart process integration tools;
2. the need to optimize the lifecycle of major projects that often change the structure of the organization implementing them;
3. multi-agent (network) decision-making with regard to the further development of the system.

It appears particularly interesting to study organizational - technical systems that form critical infrastructure. The functioning of such system involves a large number of various interactions that are determined by such properties as unpredictability, capriciousness, and the need for constant change. Some studies of the authors [9–11] demonstrate that competences of individual staff members (managers, engineers, analysts, information security experts) or of interdisciplinary teams that utilize unique tools play the key role in managing the development of such systems. It is such teams that drive structural change in OTSs making them more flexible and reliable.

2 PROBLEM OF RESILIENCE AND AGILITY IN CONTEMPORARY SYSTEMS ENGINEERING

As systems grow bigger, more complex and have a bigger environmental impact, they are required to meet tougher safety and reliability standards in changing conditions. All too often, faults in the operation of complex systems are of systemic kind. A branch of systems engineering that is concerned with building resilience systems – resilience engineering – employs methods that help avoid disasters by taking preemptive steps and increasing the adaptability and resilience of the system (its recovery after a failure).

The scholarly and engineering communities started to actively discuss the problem of systems resilience after the publication of [12]. Today, various groups of scholars look at resilience in a broader context, bringing such domains as ecology, sociology, psychology, management, engineering into the debate. The principles of resilient systems are starting to be used by risk managers and cyber security experts [13–15]. As the term "resilience" is getting widely used in various fields of science, for the purpose of this research we are going to use the definition that is accepted in systems engineering:

Resilience is the ability of a system to continue to have the required functionality in adverse conditions [16]. A typical taxonomy of resilience systems is shown in Fig. 1.

The scientific and engineering communities agree on the attributes of resilient systems. A resilient system has such features as capacity, flexibility, tolerance, and cohesion [12, 17]. Capacity means the ability of the system to survive in adverse conditions; flexibility is the ability of the system to adapt to a threat; tolerance the ability of the system to avoid a drastic loss of functionality; cohesion is the ability of the system to act as a unified whole in the face of a threat.

Principles of the architecture of resilient systems [12, 17–27] have been developed in order to endow new systems with the above attributes. Table 1 classifies and details some of these principles.

In the case of complex organizational-technical systems (for example, critical infrastructure systems) it is sufficient to ensure their resilience. The ability of a system to function in a changing environment should be factored in at the design stage. High costs and risks of reduced functionality make it irrational to apply this principle to the entire system, but it is highly reasonable to do that with regard to individual components of the system structure.

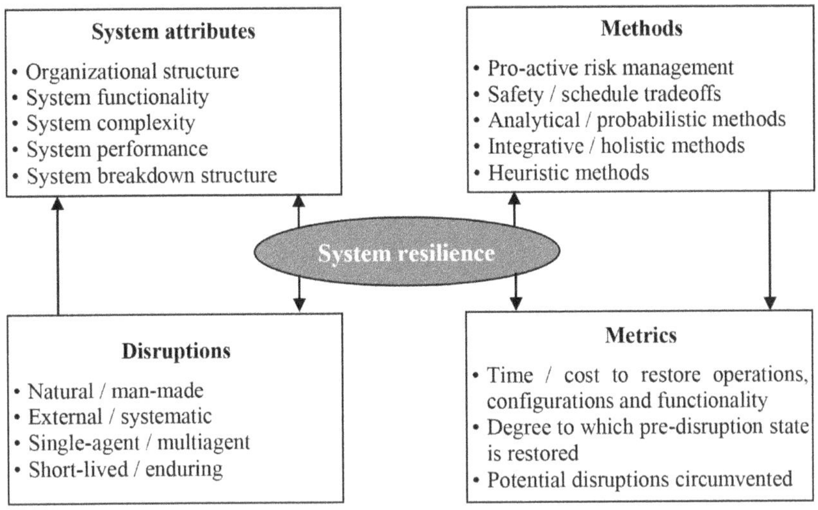

Figure 1: Conceptual framework for system resilience (adapted from [18]).

Table 1: Principles of resilient system design.

Basic principle	Definition	Supporting principles of the second level
Absorption	The system should be capable of withstanding destructive forces of a certain nature and degree	• Margin is the design security level that must exceed the supposed level of destructive impacts • Hardening means the system should have sufficient resistance to deformation • Context spanning means the system architecture should envisage the maximum destruction level as well as a wide scope of destruction • Limit degradation means that the absorption capability should prevent the system from losing functionality due to aging or poor maintenance
Restructuring	The system should be able to change its structure	• Authority escalation means that authority to manage crises shall escalate in accordance with the severity of the crisis • Regroup means that the system must restructure itself if a threat becomes real
Cross-scale interaction	Each of the system components should be capable of communication, cooperation and interaction with any other component	Knowledge between nodes means all components of the system must know what the others are doing Human monitoring implies that an automated system must understand the intent of the operator Automated system monitoring means that a human must understand the intent of the automated system Intent awareness means that all nodes of a system must understand the intent of other units

(*Continued*)

Table 1: (*Continued*)

Basic principle	Definition	Supporting principles of the second level
		• Informed operator a human must be kept informed about all aspects of the automated system • Internode impediment means that there should not be administrative or technical obstacle to the interactions among elements of a system.
Human in the loop	There should always be human in the system when the need occurs for human cognition	• Automated function means it is preferable for humans to perform a function rather than automated systems when conditions are acceptable • Reduce Human Error –standard strategies should be used to reduce human error • Human in Control- humans should have final decision making authority unless conditions preclude it
Modularity	The functionality of the system should distributed among its nodes in such a way that the failure of one of them does not bring the other ones to a halt	N/A
Neutral state	Human agents should delay in taking action to make a more reasoned judgement as to what the best action must be	N/A

Among the factors of the environment that necessitate designing so-called agile systems are capriciousness, uncertainty, risk, variation, and evolution. Taken together, they form the acronym CURVE.

The agile architecture pattern (AAP) incorporates three critical components: (1) a roster of drag-and-drop encapsulated modules that enable the designed functionality; (2) a passive infrastructure of minimal but sufficient rules and standards pull the modules together; and (3) an active infrastructure that designates specific responsibilities for sustaining agile operational capability.

For the system being developed to be able to rapidly and effectively respond to changes in the environment, it should be both proactive and reactive [28].

Proactive responses are usually triggered from within and are aimed at applying new knowledge (or implementing new opportunities for already accumulated knowledge) in order to create new value. For sustaining proactive change, the system should be capable of the following:

- module mix evolution: new modules with required functional properties are added, and inadequate modules are removed;

- improved performance of the modules in assembly thanks to their improved functionality;
- infrastructure improvement in line with evolving needs and opportunities;
- new configurations of individual modules and the roster of available modules in response to new situations that require something different in capability.

Reactive changes are generally triggered by events which demand a response. That could be a new opportunity that needs to be addressed, or a threat that needs to be countered. Reactive responses are usually associated with the properties of the systems that determine their resilience and sustainability. To respond reactively, the system should be capable of:

- correction in order to decrease the likelihood of a dysfunction and reduce recovery time if the dysfunction occurs;
- variation that is provided by module availability for an adequate response to change;
- expansion through the inclusion of new modules on the roster and use of external resources;
- reconfiguration through the employment of new modules and removal of the unused ones as well as by recycling previously disconnected modules.

3 NEW REQUIREMENTS FOR HUMAN RESOURCES

Today, management models are undergoing transformation under the influence of various global trends, With regard to the subject matter, the following ones need to be highlighted.

Super dynamic transformations in technology renewal and transfer in the global space. The ability to navigate quickly in the changing reality, foresee changes, embrace constant change has become a decisive competitive edge for any business. In this context, the goal of smartization in management is to increase the preparedness for *anticipatory action*, which means being flexible and being able to adapt to new conditions faster than your competitors by seeing new opportunities in the situation. That results in considerable shifts in managerial decision-making algorithms toward a higher level of smartization.

Broader system relationships between markets, industries, businesses and extra complex technologies and the digital environment. The trend can be easily observed in all hi-tech research-intensive industries. The growing complexity of emerging technological systems, the penetration of AI into every link of the technological chain, customized production and customer relations on the basis of cyber-physical systems, the use of virtual structures for international cooperation call for new tools and management methods [29]. Increasingly often, managers are faced with the task of promoting the newest technology in domestic and international markets, making decisions about the prospects of their application with regard to their technical characteristics. Managers need to be able to speak "the same language" with developers and producers of new equipment and operations personnel, to understand the specific features of ecosystem marketplaces, research-intensive services and their practical applications [30–32].

Past experience is being pushed off the throne by the ability to view an emerging and rapidly evolving situation *holistically*. Moreover, it becomes essential for a manager to be able to foresee looming changes. In order to be able to do that, they need to be proficient in methodology, possess strategic thinking skills, be handy with analytical support tools and instruments for designing the future. They should be able to see interdisciplinary connections, especially between the drivers of technological change.

It is obvious that a new generation of managers will be engaged in breakthroughs in technology, create new markets and overhaul existing production facilities. The key changes will be concerned with considerable improvements to productivity on the basis of smart systems, a diversity of economic and marketing for customer relations, the adoption of systems that guarantee the safety, reliability, sustainability and energy efficiency of production. For their part, major technological innovations enable profound organizational transformations, the arrival of new principles of maintenance and repair services for industrial facilities, and optimization of their life cycle.

Managers are in for qualitatively new tasks, such as:

- identification of global trends and ensuing threats and opportunities and quick adaptation to evolving circumstances;
- analysis and foresight in application to global, domestic and regional markets for technology, capital, knowledge, and competences;
- design of complex adaptive systems with innovative properties and management of their life cycle;
- finding optimal solutions factoring in multiple risks;
- management of big international projects involving virtual teams;
- interaction with experts from various domains.

It has to be noted that it is technology in the broadest sense – from targeted scientific research and engineering development projects to innovation implementation – that is a bundle of interdisciplinary connections [33, 34]. This makes it essential for an effective manager to be knowledgeable about the engineering basics of production and trends in science and technology. The authors believe that industries with super-complex technologies that can be potentially hazardous to the global eco-system (critical infrastructure industries), such as power engineering, the nuclear and aerospace industries, petrochemical production, transport, engineering and telecommunication infrastructure in cities require that managers first master the most complex interdisciplinary connections between equipment, economy, environment, and the human factor. They need to acquire versatile professional knowledge and be aware of the technical and technological specifics of the industry, the unprecedented responsibility and role of the industry in the economy.

The electric power industry serves as an illustrative example. The electricity and capacity market operates according to a certain algorithm with fairly rigid dependence on complex technological features of electric power production and mode restrictions. The production plan is determined by the operational modes of generating and grid capacity, the place of a power plan in the load schedule of the energy system and instructions issued by operational dispatch units. When considering the issue of cost and pricing, factors must be taken into account that influence the efficiency of generators and operating modes of power plants in the grid. The reliability of power supply has the priority over financial efficiency. It must comply with standards imposed by regulatory authorities and is controlled by the grid operator that maintains operating modes.

The main tools employed for *strategic management* in power engineering are rather special, too: creation and maintenance of strategic capacity reserve, flexibility of generation to eliminated uncertainties of the electrical load schedule; the energy company's policy on renewal of capital assets, demand-side management and reliability. The introduction of the Smart Grid technology makes the scientific and technological and engineering side of electrical power production and its impact on the effectiveness and timeliness of managerial decisions far more complex.

The uncertainty of the environment which energy companies operate in has grown unprecedentedly in general (financial and political crises; man-made disasters, an unstable economic situation of major energy consumers) and shows

1. the probabilistic nature of demand for electrical power and capacity temporally and region-wise;
2. instability of the stock market and inability of the banking system to provide loans to the electrical power sector due to the lack of long-term funds;
3. constantly changing tariff and energy market regulations;
4. irregular dynamics of domestic prices of fuel, natural gas above all;
5. uncertainty in the energy equipment market and about the cost of construction of major energy facilities.

It is also important that an energy company has to adequately respond to numerous regulatory activities performed by industry oversight agencies as part of the national energy policy (regional energy policy). For example, this concerns long-term capacity reserves and the expansion of electrical connections between grids in order to increase the system reliability of power supply. It is also necessary to build customized peaking power plants in specific grids to maintain the capacity balance as load schedules are expected to become increasingly uneven in regions because of an accelerated growth in the number of non-industrial consumers with peaky consumption patterns. Another important task is to develop small-scale (distributed) generation in regions, which would help offset the economic inertia of the "big" electrical power industry. It is also crucial to promote demand-side management programs as they introduce systemic changes to the energy company's relations with consumers.

Similar examples of cascades of interdisciplinary challenges for industries related to critical infrastructures make it necessary to create special mechanisms to ensure the sustainable development of complex OTSs, which, on the one hand, allow systems to maintain operational reliability, and, on the other, improve their adaptation to a changing external environment.

4 RESULTS AND DISCUSSION

Efforts aimed at designing and constructing systems and keeping them operational require teamwork that is aimed at reaching common goals and is based on holistic vision. In order to effectively perform systems engineering activities, it is necessary to make sure that the capabilities and dynamics of the team match the tasks being handled.

A systems engineering team can consist of professionals who each is responsible for a specific technical process (a requirement engineer, a systems architect etc.). It is also possible to form teams that maintain individual subsystems and are coordinated by a central team. Ideally, the team format should correspond to the distribution of roles, responsibilities, and authority in the project. The roles of the team members are determined by the structure that is in line with the strategy of the organization.

We shall illustrate this with a case from the authors' experience of forming interdisciplinary teams for the technological overhaul of a complex OTS – a large energy company that incorporates various power plants (a condensation power plant, a CHP, an HPP), several regional heating supply companies and its own engineering and technical center. The installed electrical capacity of the company is 1,256 MW; installed heating capacity is 5,886 Gcal/h. The type and scale of the company's business make it possible to classify it as a critical infrastructure system.

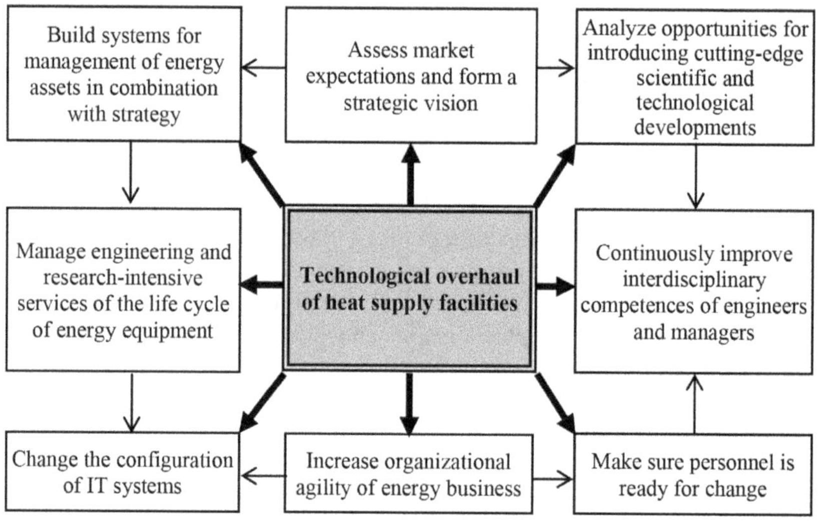

Figure 2: Tasks of organizational overhaul of OTS.

Figure 2 shows the logic of the tasks that emerge in the course of a technological overhaul. The tasks have multiple aspects to them and arise from the need to increase the agility of the OTS, which would subsequently boost modernization processes and help overcome the inertia that is typical of "big" entities in the energy sector. At the same time, the system needs to become more agile, for example, in terms of its capability to restructure itself and better interactions among the components of the OTS. That needs to be done for meeting the priority requirement for a reliable and uninterrupted power and heat supply.

In other words:

- the principles of resilience enable the organizational-technical system (the energy company in question) to stay viable;
- the principles of agility applied to individual elements of the system trigger modernization process in individual units and ensure the development of the system as a whole.

Individual processes (business areas) within the company that are connected with the technological overhaul of heat generation facilities and networks are subject to the CURVE factors defined in Table 2.

In order to mitigate the above factors, a decision was made to set up so-called interdisciplinary breakthrough teams within the company that would consist of professionals in various domains. By engaging the teams in strategy-building activities, the development and testing of forward-looking technical and economic solutions that corresponded to the vision, it was possible to present three innovative projects to the company management within a fairly short period of time (18 months). The projects were aimed at new market exploration, renovation of capital assets and launch of new production facilities, reform of the corporate culture and a transition to the professional development of the personnel. Interdisciplinary teams address problems that cannot be solved within the existing paradigm (internal architecture of the company). That's why their task is to change the traditional approach to systemic development and business processes. aT the same time, the members of the team continue to head

Table 2: CURVE-factors of environment in application to the studied OTS.

Factor		Characteristics	
Proactive		Reactive	
C	**C**apriciousness: unknowable situations	Unitended consequences of the use of new technical solutions; personnel training and adaptation programs are future-oriented; require change in established practice and envisage interactions with customers who have specific demands for energy supply	Energy efficiency programs implemented by consumers; possible establishment of free zones in the area whose tenants have unconventional requirements for electricity supply; proficiency of young specialist lags behind market demands
U	**U**ncertainty: randomness with unknowable probabilities	Uncertain compatibility (behavior) of equipment and automation solutions of different types and by different producers	Major consumers increase and decrease consumption depending on market situation; development of on-site and small-scale generation
R	**R**isk: randomness with knowable probabilities	Technological development of facilities (technical overhaul plans), launch and decommission of capacity, staff training contracts with training contractors	Electricity and gas tariff regulation; impact of climate on consumption
V	**V**ariation: knowable variables and associated variance ranges	Testing, fine-tuning and utilization of new technological solutions at existing facilities, different proficiency levels	Facilities with different service life, operational parameters and specifications
E	**E**volution: gradual successive developments	Development and implementation of corporate development strategy (including, technology, social, environmental and other aspects)	Launch of new facilities equipped with cutting-edge equipment, upgrades to machines during major repairs, natural process of recruitment, dismissal and reemployment of staff

their units and promote ideas to their co-workers. So-called growth points appear. They do not destroy the system, but ensure the flexible development of its individual components.

Figure 3 depicts interactions with an Interdisciplinary Breakthrough Team (iTeam). The figure was constructed on the bases of architectural patterns of agile systems.

Breakthrough teams boast the following competences (capabilities).

1. They see what happens beyond the system boundaries and understand how it will affect it.
2. The can estimate the feasibility of transformations (considering resources, intellectual potential, time) and select instruments (means) for executing them.

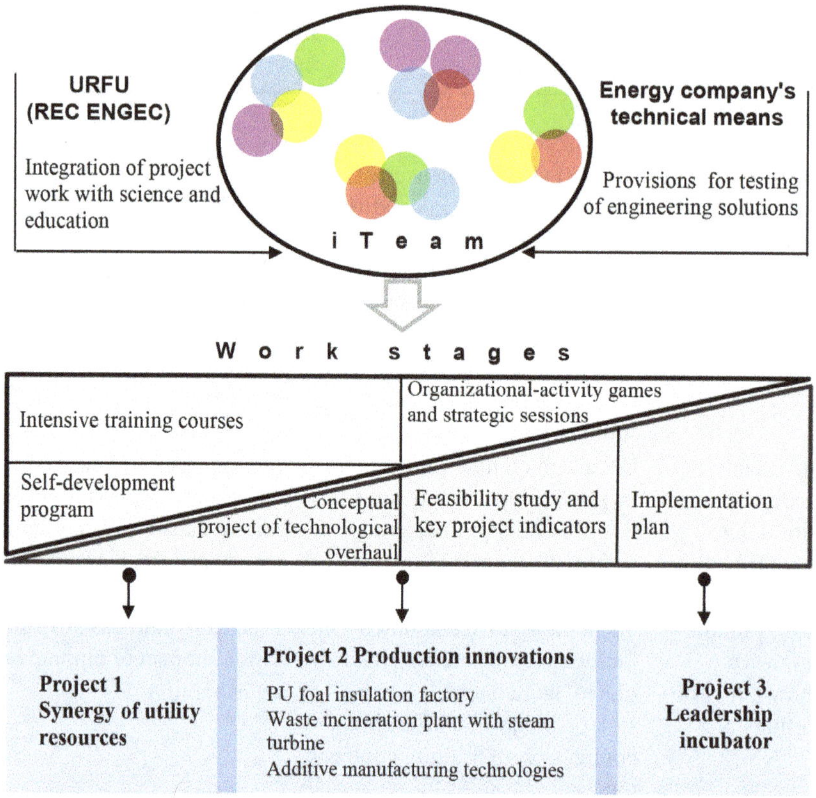

Figure 3: Formation of interdisciplinary teams for technological overhaul.
Note: The color circles indicate that the team includes professions in different domains.

3. They realize what transformations need to happen in the mindset of staff members and identify growth points in the process.
4. They have conceptual design skills.

In general, it was possible to make the OTS in question more agile by

- engaging the management in a discussion of the initiatives and securing the top managers' support for the teams;
- carefully selecting members of the teams, who were subjected to a series of diagnostic procedures and special procedures that boost teamwork at different stages;
- developing simultaneously engineering and organizational solutions, which made it possible to minimize transformation risks in the company;
- active participation of REC ENGEC consultants as moderators, who provided a holistic vision of technological overhaul tasks and of the end result.

5 CONCLUSION

Elements of critical infrastructure must be highly resilient and, at the same time, be capable of development and rapid response to operation conditions. By sticking to the principles of resilient systems design, it is possible to maintain the required functionality in adverse conditions and add components to the system that ensure its development and transformation.

The most acute problem in the implementation of the latter possibility stems from the inertia of an organizational system and the difficulty of striking the balance between the required resilience and agility. The problem cannot be handled without serious organizational transformations that have to be conducted by interdisciplinary teams.

The method of agile systems-engineering that was developed as part of systems engineering could be used for assembling and training such teams. However, the method needs some adjustment due to the specific goals that breakthrough teams are faced with and their conditions they operate in.

Research into the training of such teams for a major energy company made it possible to identify critical competences (capabilities) that such a team need to possess.

The mechanism of integrating such teams into the system of critical infrastructure facilities need further elaboration so that the potential of such teams could be exploited to the full and the balance between the resilience and agility of the system in general is maintained.

ACKNOWLEDGEMENTS

The work was supported by Act 211 of the Government of the Russian Federation, contract No. 02.A03.21.0006.

REFERENCES

[1] Trist, E.L., Higgin, G.W., Murray, H. & Pollock, A.B., *Organizational Choice: Capabilities of Groups at the Coal Face Under Changing Technologies*, London: Tavistock Publications, 1963, 332 p.

[2] Burlankov, S.P., Burlankov, P.S. & Skvortsov, A.O., Basic concepts and classification of socio-technical, techno-economic and socio-economic systems [in Russian]. *University Proceedings. Volga Region. Social Science*, **42(2)**, pp. 169–183, 2017. https://doi.org/10.21685/2072-3016-2017-2-17

[3] Ladaniuk, A., Prokopenko, T. & Reshetiuk, V., The model of strategic management of organizational and technical systems, taking into account risk-based cognitive approach. *Annals of Warsaw University of Life Sciences – SGGW, Agricultural and Forest Engineering*, **63**, pp. 97–104, 2014.

[4] Baxter, G. & Sommerville, I., Socio-technical systems: From design methods to systems engineering. *Interacting with Computers*, **23(1)**, pp. 4–17, 2011. https://doi.org/10.1016/j.intcom.2010.07.003

[5] Kornakov, A.N., Model of complex organizational-technical system. *International Scientific Electronic Journal*, **2(14)**, pp. 44–50, 2015.

[6] Burkov, V.M., Korgin, N.A. & Novikov, D.A., Control mechanisms for organizational-technical systems: problems of integration and decomposition. *IFAC-PapersOnLine*, **49(32)**, pp. 1–6, 2016. https://doi.org/10.1016/j.ifacol.2016.12.180

[7] Teslenko, P.A., Evolutionary development of organizational and technical systems [in Russian]. *Eastern European Journal of Advanced Technology*, **4(48)**, pp. 22–25, 2010.

[8] Shrum, W. & Robert, W., Reputational Status of Organizations in Technical Systems. *American Journal of Sociology*, **4(93)**, pp. 882–912, 1988. https://doi.org/10.1086/228828

[9] Gitelman, L.D., Sandler, D.G., Gavrilova, T.B. & Kozhevnikov, M.V., Complex systems management competency for technology modernization. *International Journal of Design & Nature and Ecodynamics*, **12(4)**, pp. 525–537, 2017.
https://doi.org/10.2495/dne-v12-n4-525-537

[10] Gitelman, L. & Kozhevnikov, M., New leaders for technological breakthrough in the energy industry. *WIT Transactions on Ecology and The Environment*, **224**, pp. 499–511, 2018.

[11] Gitelman, L.D., Gitelman, L.M. & Kozhevnikov, M.V., Managers for sustainable electric power industry of tomorrow. *International Journal of Sustainable Development and Planning*, **13(2)**, pp. 307–315, 2018.
https://doi.org/10.2495/sdp-v13-n2-307-315

[12] Hollnagel, E., Woods, D.D. & Leveson, N., *Resilience Engineering: Concepts and Precepts*, Aldershot: Ashgate Publishing Limited, 2006, 397 p.

[13] Connelly, E.B., Allen, C.R., Hatfield, K., Palma-Oliveira, J.M., Woods, D.D. & Linkov, I., Features of resilience, *Environment Systems and Decisions,* **37(1)**, pp. 46–50, 2017.
https://doi.org/10.1007/s10669-017-9634-9

[14] Bodeau, D.J., Graubart, R.D., Picciotto, J. & McQuaid, R., Cyber Resiliency Engineering Framework, 2011, available at https://www.mitre.org/sites/default/files/pdf/11_4436.pdf (accessed 12 March 2018)

[15] Bodeau, D. & Graubart, R., Cyber resiliency design principles. Selective use throughout the lifecycle and in conjunction with related disciplines, 2017, available at https://www.mitre.org/sites/default/files/publications/PR%2017-0103%20Cyber%20Resiliency%20Design%20Principles%20MTR17001.pdf (accessed 12 March 2018).

[16] Walden, D.D., Roedler, G.J., Forsberg, K.J., Hamelin, R.D. & Shortell, T.M., *INCOSE Systems Engineering Handbook: A Guide for System Life Cycle Processes and Activities, 4th Edition*, Hoboken, New Jersey: Wiley, 2015, 304 p.

[17] Jackson, S. & Ferris, T., Resilience principles for engineered systems. *Systems Engineering*, **16(2)**, pp. 152–164, 2013.
https://doi.org/10.1002/sys.21228

[18] Madni, A., & Jackson, S., Towards a conceptual framework for resilience engineering. *Institute of Electrical and Electronics Engineers (IEEE) Systems Journal*, **3(2)**, pp. 181–191, 2009.
https://doi.org/10.1109/jsyst.2009.2017397

[19] Richards, M.G., Multi-Attribute Tradespace Exploration for Survivability, 2009, available at http://citeseerx.ist.psu.edu/viewdoc/download?doi=10.1.1.207.8920&rep=rep1&type=pdf (accessed 12 March 2018)

[20] Maxwell, J. & Emerson, R., Observations of the resilience architecture of the firefighting and emergency response infrastructure. *Insight*, **12(2)**, pp. 45–47, 2009.
https://doi.org/10.1002/inst.200912245a

[21] Jackson, S., *Architecting Resilient Systems: Accident Avoidance and Survival and Recovery from Disruptions*, Hoboken, New Jersey: Wiley, 2010, 307 p.

[22] Haddon-Cave, C., An independent review of the broader issues surrounding the loss of the RAF Nimrod MR2 Aircraft XV230 in Afghanistan in 2006, available at http://news.bbc.co.uk/2/shared/bsp/hi/pdfs/28_10_09_nimrod.pdf (accessed 12 March 2018)

[23] Billings, C.E., *Aviation Automation: The Search for Human-Centered Approach,* Mahwah, NJ: Lawrence Erlbaum Associates, 1997, 355 p.

[24] Perrow, C., *Normal Accidents: Living With High-Risk Technologies*, Princeton, NJ: Princeton University Press, 1999, 464 p.
[25] Leveson, N., *Safeware: System Safety and Computers*, Massachusetts: Addison-Wesley, 1995, 680 p.
[26] Reason, J., *Managing the Risks of Organisational Accidents*, Aldershot, UK: Ashgate Publishing Limited, 1997, 252 p.
[27] Reason, J., *Human Error*, Cambridge: Cambridge University Press, 1990, 302 p.
[28] Dove, R., Fundamental Principles for Agile Systems Engineering, 2005, available at www.parshift.com/Files/PsiDocs/Rkd050324CserPaper.pdf (accessed 12 March 2018).
[29] Delphi Energy Future 2040. Delphi-study on the future of energy systems in Germany, Europe and the world by the year 2040, available at http://www.pwc.com/gx/en/energy-utilities-mining/pdf/delphi-energy-future.pdf (accessed 12 March 2018).
[30] Schönmann, A., Greitemann J. & Reinhart G., Proactive management of production technologies: a conceptual framework. *Procedia CIRP*, **52**, pp. 198–203, 2016. https://doi.org/10.1016/j.procir.2016.08.011
[31] Scholten, D. & Künneke, R., Towards the comprehensive design of energy infrastructures. *Sustainability*, **8**, p. 1291, 2016. https://doi.org/10.3390/su8121291
[32] Giret, A., Garcia, E. & Botti V., An engineering framework for Service-Oriented Intelligent Manufacturing Systems. *Computers in Industry*, **81**, pp. 116–127, 2016. https://doi.org/10.1016/j.compind.2016.02.002
[33] Schoemaker, P.J.H., Day, G.S. & Snyder, S.A., Integrating organizational networks, weak signals, strategic radars and scenario planning. *Technological Forecasting & Social Change*, **80**, pp. 815–824, 2013. https://doi.org/10.1016/j.techfore.2012.10.020
[34] Chryssolouris, G., Mavrikios, D. & Mourtzis, D., Manufacturing systems: skills & competencies for the future. *Procedia CIRP*, **7**, pp. 17–24, 2013. https://doi.org/10.1016/j.procir.2013.05.004

COEVOLUTION OF TECHNOLOGY, BUSINESS AND SOCIETY

GEORGE RZEVSKI
The Open University and Digital Ecosystems Ltd.

ABSTRACT
Social changes created during technological paradigm shifts in the past are reviewed with a view to establishing patterns of changes that can help in forecasting the social change that are likely to follow the current technological revolution.
Keywords: Coevolution, Complexity, Emergence, Selforganisation.

1 INTRODUCTION

Society invests into entrepreneurs and inventors who create new technology, which, in turn, induces structural changes into economy by creating demand for new jobs that require new skills, and in the process, destroying the old employment opportunities.

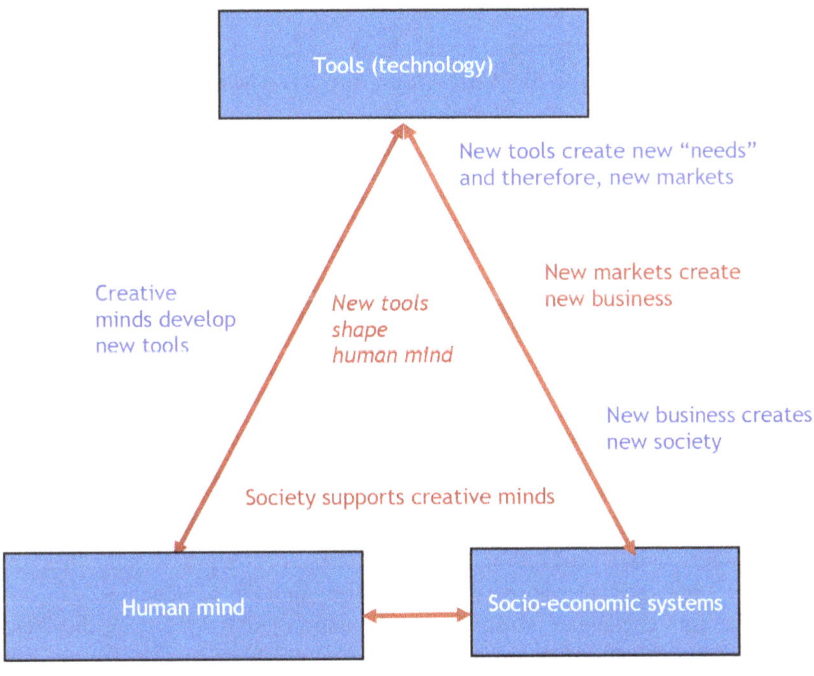

Figure 1: Interplay of creativity, technology and society.

2 COEVOLUTION: MAJOR STAGES

Historically, complexity of social systems increased in three very large steps, driven by the advances in technology, as depicted in Table 1. At each step, the impact on the way we lived and worked was dramatic – leading entrepreneurs and members of society who acquired new skills emerged as the new social elite.

DOI: 10.2495/DNE-V13-N3-231-237

Table 1: Coevolution of technology, business and society.

STAGES	KEY RESOURCES	DISTRIBUTION	SCOPE	SUCCESS FACTORS
Information Society Knowledge-based services Digital technology	Knowledge	Digital Networks	Global	Adaptability
Industrial Society Production of goods Mass-production technology	Capital	Railways Motorways	National	Economy of Scale
Agricultural Society Production of food Agricultural technology	Land	Local Roads	Local	Economy of scale
Hunter-gatherers Hunting for food Simple tools	Personal skills	None	Tribal	Teamwork

Perhaps the most dramatic structural social changes occurred during the transition from the ***Society of Hunter-Gatherers*** to the ***Agricultural Society***, when the new technology for cultivating large fields enabled Homo Sapience to settle down into permanent rural communities. In the process, society changed from a large number of small groups of equals, continuously on the move, to a rich elite of landowners and a huge army of poor agricultural workers in permanent settlements.

The ownership of ***Land*** became the key class distinction, and the gap between rich and poor reached unprecedented levels [1].

During the transition from the ***Agricultural Society*** to the ***Industrial Society*** the rapid migration of the population from the countryside to the cities, to take advantage of new employment opportunities, increased the social connectivity in the densely populated cities and, as a result, raised the level of social complexity. The massive movement of population caused well-documented disturbances as a rigid, traditional social order based on land ownership was replaced by a chaotic transition, which then settled into a new social order based on ownership of ***Capital***. According to Arnold Toy1720nbee [2] "the essence of the Industrial Revolution" was "the substitution of competition for the medieval regulations which had previously controlled the production and distribution of wealth". And, "the *Wealth of Nations* and the steam-engine...destroyed the old world and built a new one."

Driven by the unprecedented advances in information technology (IT), the West is currently in the middle of the third significant socio-technological revolution, the transition from ***Industrial Society***, dominated by trade in physical goods, to ***Information Society***, characterised by widespread trading in knowledge-based services [3].

The rapid expansion of information technology is currently producing an unprecedented increase in complexity of markets [4], a drastic shift in employment opportunities and will, in time, bring important social changes. The research reported here was set to investigate direction of these changes.

In economic terms, the change is dramatic. ***Knowledge*** is replacing ***Capital*** as the main business resource [5]. The UK is a good example; whilst after the Second World War

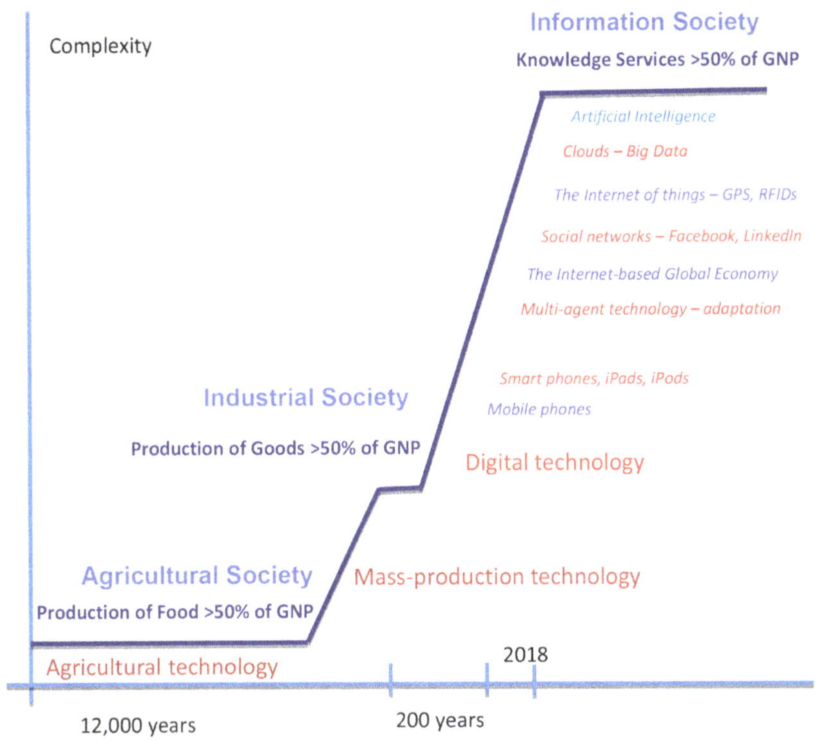

Figure 2: Complexity of technology, business and society increases at every mayor paradigm shift.

manufacturing generated around 50% of the UK GDP, its current contribution is below 15%. At present, the prime wealth generating engine is the service sector, which contributes over 75% to the UK GDP [6].

The US is leading the West into a brave new world of knowledge economy, as is evidenced by the new business elite – Apple, Google, Amazon, Facebook, YouTube and Twitter, which have pushed the car and oil giants (remnants from the industrial era) onto the B list and have propelled IT pioneers, like Bill Gates, to the top of the rich list.

3 THE POWER OF DIGITAL TECHNOLOGY

3.1 Phase 1 – Digital Computation

The early beginnings of digital technology focused on increasing computational power of computers. The leaders, companies such as IBM, built larger and larger mainframe computers.

3.2 Phase 2 – Digital Communication

Then came a transitional period focused on developing networks of smaller computers, thus combining computation with communication. The new technology of servers and workstations enabled new, innovative companies, such as Sun Microsystems and Microsoft, to rise to the top, forcing the previous generation of leaders to rethink their strategy.

Later on, the second phase almost completely focused on digital communication technology. Apple achieved prominence by supplying consumers with advanced digital communication devices such as smart phones and tablets, and by offering services such as data streaming.

3.3 Phase 3 – Digital Intelligence

The new phase is all about artificial intelligence (AI). Advances in AI are leading to a steep increase in the number of autonomous robots and driverless vehicles capable of performing better the tasks previously performed by human operators.

Much more importantly, intelligent software, capable of performing clerical, professional and managerial tasks more effectively than humans, will take over from human decision makers leaving them free to take up new, previously unknown, employment opportunities.

4 NEW BUSINESS

Digital communication technology increased social and economic ***connectivity*** and, as a consequence, socio-economic ***complexity*** reached new levels.

The Internet-based global market has become highly complex with billions of individuals and millions of businesses engaged in setting up, modifying, and/or cancelling business transactions with each other at great speeds and without geographical constraints. The volatility and unpredictability of demand and supply began to affect businesses and by now, it is well understood that businesses must develop capability to rapidly ***adapt*** to any disruptive event generated by markets. And since the frequency of disruptions is high, the decisions related to adaptability must be performed with speeds with which human decision makers cannot cope.

It follows that business must replace a large number of human operational decision makers with AI.

How Market Complexity Affects Business

- Unpredictable and volatile demand
- Ever-changing competition
- Increased frequency of disruptive events such as:
  - Arrival of unexpected orders (claims)
  - Non-arrival of expected orders (claims)
  - Changes to accepted orders (claims)
  - Delays
  - Failures
- Increased frequency of incidents:
  - Fraud
  - Hacking
  - Electronic attacks
- Planning deliveries difficult, if not impossible
- Traditional resource scheduling methods obsolete

The following business processes (and many others) will be affected:

- supply chains,
- production
- purchasing
- order processing,
- distribution
- warehousing
- transportation
- water supply
- food supply
- project management

We are facing an interesting paradox: complexity, caused by digital communication can be manged only by employing digital intelligence. In other words, digital technology is responsible for an issue, which can be resolved only by employing digital technology.

5 CASE STUDY: NEW INSURANCE

A research conducted by the author in cooperation with a large insurance company revealed feasibility of converting a significant number of insurance-related business processes into autonomous mode, in which all operational decisions are made by AI systems. Few examples are given below.

- Extracting knowledge from data on individual clients
- Creating individual polices in interaction with individual clients
- Extracting knowledge from data on competition
- Dynamic risk assessment driven by risk affecting events
- Claim processing performed by AI without human involvement

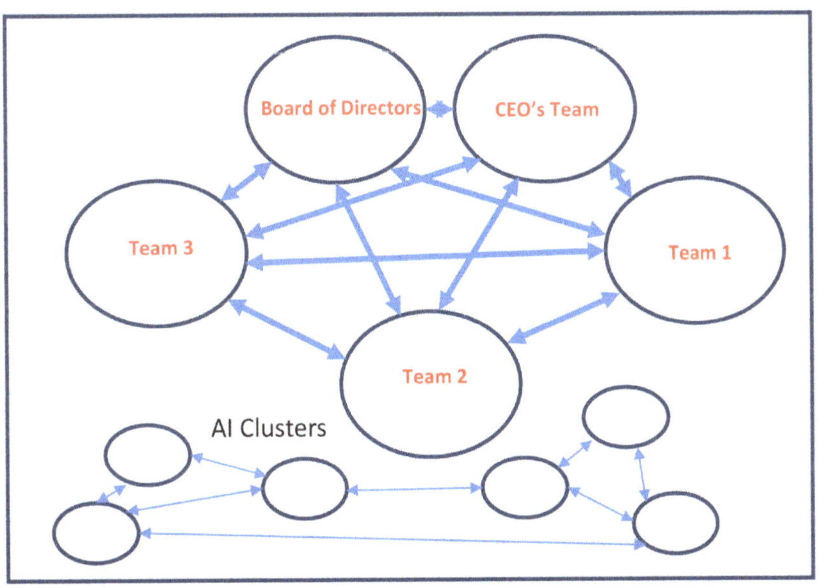

Figure 3: Cooperation of human and digital intelligence.

Human intelligence, working in cooperation with digital intelligence, would be responsible for generating and selecting strategic options, innovating, negotiating, advising, supervising, inspecting and servicing.

A prototype AI system is currently being developed to explore autonomous insurance claim management.

6 DIGITAL TECHNOLOGY AND SOCIETY

During the last two major technological paradigm shifts, we have experienced momentous social and political change. First time, when simple tools that were used by hunter-gatherers, were replaced by extensive agricultural technology, which enabled ever-moving tribes to settle into permanent residences; and second time, as agricultural technology was replaced by mass-production technology as a main economic driving force.

Can we expect less significant changes when, finally, digital technology completes the shift from industrial era, with its large, rigid corporations and the economy of scale, to more nimble, adaptive information economy? I doubt it.

7 NEW ESTABLISHMENT

We shall probably witness the demise of the current establishment and the emergence of socially-aware knowledge workers, such as Bill Gates, ready to spend huge sums of money accumulated during working life, on wellbeing of the disadvantaged.

Table 2: Transformation of Business & Society.

	Society of Hunters and Gatherers	**Agricultural Society**	**Industrial Society**	**Information Society**
Main economic activity	Providing resources for family/tribe	Production of food	Manufacturing of goods	Knowledge-based services
Main technology	Simple tools	Agricultural technology	Mass-production technology	Digital technology
Main resource	Personal skills	Land	Capital	Knowledge
Key success factor	Personal skills	Economy of scale	Economy of scale	Adaptability
Distribution	None	By roads	By railways and motorways	By digital networks
Scope	Family/tribe	Local	National	Global
Dominant class	None	Landowners	Bankers, lawyers, industrialists	Knowledge workers
Management	Coordination	Unconstrained command	Corporate hierarchies Command & control	Distributed decisions Use of artificial intelligence
Governance/ politics	Wisdom of tribal elders	Feudal	Party-centred democracy	Digitally managed issue-centred democracy

Current predominance of bankers and lawyers in all walks of life is likely to disappear.

8 NEW DEMOCRACY

Democracy based on antagonistic political parties led by professional politicians, without any practical experience outside party politics, is likely to be replaced by issue-based democracy, where each important problem attracts attention of politically minded citizens who are prepared to pull their resources and, supported by powerful digital technology, collectively work on achieving the resolution.

9 CONCLUSIONS

As father of Complexity Science, Ilya Prigogine, stated [7]: "The future is not given". It is perpetually constructed by decisions and actions of agents active in the known Universe. And yet, it is not random. It evolves following discernible patterns [8].

By understanding patterns of socio-economic changes that were caused by technological paradigm switches in the past, we can reasonably predict the impact of digital technology on business and society in the near future.

REFERENCES

[1] Harari, Y.N., *Sapiens – A Brief History of Humankind,* Harvill Secker: London, 2014.

[2] Toynbee, A., *Lectures on the Industrial Revolution in England*, Rivington's: London, 1884.

[3] Rzevski, G. & Skobelev, P., *Managing Complexity*, WIT Press: Southampton, Boston, 2014.

[4] Rzevski, G., Complexity as the defining feature of the 21st Century. *International Journal of Design & Nature and Ecodynamics*, **10(3)**, pp. 191–198, 2015. https://doi.org/10.2495/dne-v10-n3-191-198

[5] Drucker, P.F., *The Age of Discontinuity*, Harper & Row: New York, 1992.

[6] Wikipedia, the UK GNP.

[7] Prigogine, I., *Is Future Given?* World Scientific Publishing Co.,Singapore, 2003.

[8] Prigogine, I., *The End of Certainty: Time, Chaos and the New Laws of Nature*. Free Press: New York, 1997.

ON PEOPLE AND COMPLEXITY IN HEALTHCARE SERVICE SUPPLY

PER ENGELSETH[1] & BRIAN E. WHITE[2]
[1]*Department of Logistics, Molde University College, Norway.*
[2]Complexity Are Us ← Systems Engineering Strategies, USA.

ABSTRACT
Healthcare logistics is treated as a fundamentally emergent complex system primarily because a broad range of stakeholders are included. The patient is the primary "customer" of the service producers in the supply chain, including the doctors, nurses, medicine and insurance providers, and hospital administrators. Problematic issues regarding healthcare quality that need to be solved or improved are discussed, and suggestions for furthering and accelerating progress are offered. Careful application of information technology in designing appropriate information systems is advocated. Three specific illustrative cases of healthcare services that have been analyzed and assessed are summarized. The overall intent is to motivate creative processes for delivering more efficient and effective healthcare utilizing complex system behaviors and engineering principles, and an ethically-founded worldview.
Keywords: case studies, collaboration, complex systems engineering, complex systems, ecosystems, healthcare services, information systems, interdependencies, logistics, process emergence, supply chain management

1 INTRODUCTION

Healthcare logistical support is crucial in enabling overall service quality. Clearly, there are shortcomings in this and other areas, especially in the US, where healthcare expenditures represent about 20% of the Gross National Product (GNP). In addressing these issues here, we consider people problems as the most critical but we are also interested in healthcare management, and how best to apply information technology (IT) in creating information systems (ISs) that can make a positive difference.

Primarily because of the intensive composition of people, e.g., patients and care providers, in particular, understand that healthcare services are a very complex system. One of the key tenets in dealing with a complex system is the use of stakeholder analysis to help form the basis for actions in identifying and analyzing problems, and formulating potential solutions to be explored and implemented, perhaps on a trial basis to learn whether they help or not. The interactions of people both within and outside their organizations need to be studied, characterized, and revamped, as necessary. In a supply chain producers and customers are both value-creating and sensemaking agents that impact healthcare quality.

Behaviors of complex systems, and complex systems engineering principles [1] are cited throughout to help emphasize and generalize some of the specific points and comments of this paper. The overall hope is to raise awareness of the more important issues and to help stir concern and follow-up action to improve healthcare delivery and outcomes.

2 BACKGROUND

In this section some previous and *apropos* work in the fields of complexity, systems thinking, and complex systems engineering are cited. Healthcare problems and potential pathways to solutions are also discussed.

DOI: 10.2495/DNE-V13-N3-238-249

2.1 Previous Work

The nature of complex systems and suggestions for complex systems engineering are described in many publications, e.g, [2–9]. In perusing the literature on these topics, one soon realizes that the people associated with the system often generate a high degree of complexity; and that intentionally including such stakeholders in the system to be engineered is necessary for creating more viable pathways toward system development and improvement. Nevertheless, good, available, up-to-date, and tested technology is often crucial in helping to solve human-centric complex problems. We propose to apply IT in establishing ISs amenable to effective healthcare logistical supply chain capabilities.

IT's resource base includes *integrated* information hardware and software design, development, and use. Three aspects of IS potential are discussed in [10]: (1) features of connectivity associated with people and documentation; (2) aspects of big data (BD); and (3) application of IT to integrate multi-tier service supply chains. Organizing such information, focusing on agency in a complex system, and describing the tools used by people (primarily the agents) to reach system goals seeks to produce supply chain value as the overarching system purpose. In addition, the ethical aspects of all three of these issues (connectivity, BD, and IT) should be considered. This is especially pertinent since healthcare concerns logistical services that directly impact human well-being.

2.2 What's the Problem?

Healthcare services need to be improved. Witness the prevailing indicators of exorbitant hospital, doctor, and drug costs, uninsured emergency room visits, and insufficient emphasis on preventative medicine, for example. Since people are embedded in organizing, operating, managing, and communicating within the logistical supply chain of healthcare, whatever improves the interactions of those people in providing healthcare services should positively affect overall performance. Later in this text we provide some narratives describing current actual standing of services in this industrial sector.

We deemphasize what seems to be a common mindset that technology alone can/will provide all the answers in enabling and delivering healthcare services. Adopting a people-centric "mindsight," [11] a term suggesting more open-minded flexibility than the rigidity-of-thought term, "mindset," seems more appropriate. Furthermore, viewing the healthcare providers as a highly interconnected and interactive network broadens the usual scope and expands the often too-narrow boundary of the system of interest.

Fundamentally, people are not only important in the production-and-supply side of healthcare, other supporting agents, people associated more with patients, for example, are the principal constituents of the healthcare problem side. The healthcare service providers and the related agents need to be concerned with not only: (1) how best to treat the sick efficiently; but also, better yet, (2) our overarching higher moral responsibility to make and keep everyone healthy more effectively. However, even then, individual mindsets/sights can be rather limited, and a healthy collaboration is required.

Healthcare services comprise a special industry with important particularities to grasp in order to develop them further. Interdependencies vary between different types of supply chains since they impact how people interact to produce a service. Turning now to the interactions of the people involved in the healthcare logistical supply chain, the nature of their interdependencies arises as an important issue. As discussed in detail in [10], interdependencies can range from sequential (serial), to reciprocal (two-way), to pooled (networked), where these three types are characterized as moving in the direction of increasing importance (value) in similarly more complex systems. To the extent one can design these interdependent relationships, the focus should be on *networked* interdependencies as the most general and potentially most beneficial, where little, if anything, will be omitted from consideration. However, after thorough analysis some networked interdependencies might be reduced to the reciprocal type without any loss of benefit but with a concomitant increase in effectiveness or efficiency. Similarly, upon further detailed analysis, a reciprocal interdependency might be simplified to a sequential type to achieve a net gain in effectiveness/efficiency without any loss in benefit.

This is akin to saying that classical conventional, or traditional systems engineering techniques certainly can be usefully applied if the system at hand is not too complex. However, the more difficult systems, viz., system of systems (SoSs), enterprises, and the most complex systems require complementary techniques to bring the "requisite variety" [12] necessary to cope with the problems.

Network interactions are key in producing and delivering exemplar services in almost any domain but especially in healthcare where people are central players that comprise localized and extended networks spanning their concerns, whether conscious and intentional, or unconscious and intuitive. Through their actions, assisted with state-of-the-art technology to enable effective and efficient communications, healthcare givers and recipients engender a dynamic network of nodes and links, where the more influential can be represented by larger nodes with more incoming and outgoing links, akin to a "hub and spoke" supply chain.

The customers need to provide inputs (e.g., their identity, location, tangible resources, and health concerns) to, and feedback (e.g., their updated health status, degree of satisfaction, and further requests) about the healthcare process. Inadequate service quality results from, e.g., *ad hoc*, seemingly random, and delayed or duplicate message arrivals; their unclear content or inconsistent specifications; and errors in tracking current medical inventories, producing remedies, and implementing operational transactions. For more detail, cf., [10].

Providers are typically either unaware of or distracted from these difficulties. They tend to make insufficient efforts to mount systemic studies of the problems, achieve consensus among their collaborators, and commandeer resources necessary for significant improvements. Pooling different types of resources and configuring services in different ways increase service production complexity. Hospitalization involves a great deal of variation. For instance, a given ailment may be confronted differently even within the same hospital. This variation in service includes the numbers and types of patients in the reception ward, various healthcare personnel with different operational preferences, and the use of non-standard rooms and equipment, as well as time-line variations in capacity as impacted/constrained by other patient numbers, types of ailments, room availabilities, tools, medicines, health personnel, etc. [10].

2.3 What's the Solution?

Before proceeding it is important to understand or realize that system complexity is not necessarily a random chaotic *thing* out there. The behaviors of complex systems, most effectively operating at the *edge of chaos*, are not "random" or stochastic; they do not arise purely by chance. Such a "healthy" complex system continuously (or continually) evolves based on component interactions within its boundary and interactions with its environment across its boundary. Thankfully, a complex system can be influenced by design, or, more likely, thoughtful interventions that may, with sufficient time, lead to improved behaviors—or not; if the latter, one must consider taking a different approach and try to re-intervene [6].

What can one discover about how a complex system is responding to its interactions or our interventions? Typically, these responses are exhibited by patterns that may be observed but are often difficult to interpret. Simple, direct responses cannot often be reliably correlated with particular interactions or interventions. So, although there is cause for hope, the challenges are great. However, much can be learned about the patterns of a complex system by applying the tools of agent-based modeling and simulation in a virtual imitation of the complex system that operates in parallel with the original, perhaps in a speeded-up time frame [1]. We encourage researchers and workers in healthcare to pursue such approaches to help discover improved methodologies.

Healthcare logistics, with its inherent industrial particularities, may be effectively described and analyzed, for development as well as operational purposes, as a complex system supported by an IS. Complexity can be defined as: "… a property of an open system that consists of a large number of diverse, intelligent, partially anonymous, richly interconnected components, often called agents, [with] no centralized control and whose behavior emerges from the intricate interaction of agents and is therefore uncertain without being random" [13, p. 5]. These authors, Rzevski and Skobelev, also point out that the key features of complexity include openness, diversity, partial autonomy and interconnectedness of agents, lack of centralized control, and emergence. The complex systems' components interlinking and interacting with its environment exhibit considerable interdependence.

Since emergence, the largely unexpected manifestations of observable properties, is a prominent characteristic of complex systems, we must recognize the likely existence of heightened degrees of process uncertainties. In attempting to handle these uncertainties, one is advised to bring healthy doses of humility to bear [1], and to try to instil a spirit of flexibility in those involved in service production and management operations, and information, monetary, and expertise exchanges, at all levels of our healthcare institutions or organizations.

Healthcare service development involves organizing people to act and interact effectively not only in person but also in utilizing IT. Importantly, most everyone involved in such an effort needs to adopt an open and constructive attitude of respecting and deeply listening to others with the hopes of increasing the collective understanding of the underlying realities of healthcare problems while creating appropriate bases for possible solutions. During such discussions, the group should also include attempts to clarify the boundary of the complex system to be improved, hopefully with consensus. Restricting the boundary too much may omit key elements of the problem, while expanding the boundary too much may make viable solutions nearly impossible. As Einstein has challenged us, make the problem as simple as possible but no simpler!

3 ELABORATION

From practical and ethical viewpoints healthcare service development is a global challenge. We now, however, turn to focus on the status quo of healthcare services in one of the world's most affluent nations, the US. As discussed thoroughly in [14] the US healthcare system may be terminally broken unless some more widespread and additional drastic measures are taken to further improve and correct the current situation. It is generally accepted that several European healthcare systems cost about half of what the US is expending under our current "fee-for-service" system. And one can argue whether US healthcare results are any more effective than those of the more socialized European systems, even discounting the large number of US residents that don't have access to healthcare. Here we will focus on the US.

3.1 Let's Somehow Change the Status Quo for the Better

According to [14] approximately a quarter of a million people die every year from the shortcomings of the US healthcare system. There are misdiagnoses, inappropriate treatments, mistakes in delivering the proper medication and other remedies, etc.

Medical information and health records are still not being pervasively shared between or among hospitals. This resistance is due largely to protecting hospital "businesses" by jealously hiding medical information that may be useful to others, often citing the need to protect patients' privacy. Incompatibility of data processing and interoperability equipment, software, and procedures, for example, is also a paramount issue [15].

The prevailing fee-for-service dominated practices tend to reward more care rather than the better care. More than 90% of the doctors get paid that way, and almost 80% say they don't want to change from that system. Approximately $1 trillion is wasted on treatments that patients do not really need.

Almost everyone in the system is charging "what the market will bear." In 2007, for example, lobbyists spent about $150 million to avoid changing that paradigm. They dominate in market control. There are instances of monopolistic behaviors which exacerbate cost increases such as with the recent epinephrine auto-injector EpiPen [16] fiasco.

Hospitals are investing in specialty care because that's good for the "bottom line." It appears that nothing will change the current set of incentives. Most players appear to be content with current practices and will likely resist any systemic attempts to provide more attractive cost saving measures.

Further population growth may lead to greater difficulties in funding proper education to help embolden future generations to counterbalance the current trend toward "Trumpism" in the US and reactionary nationalism in several other countries, including France, Germany, and the UK, for example. If the lower and middle classes, as well as women in general, continue to earn relatively less on almost any comparison scale, as robots take over more and more jobs, putting people out of work and making normal jobs and reeducation more difficult to attain [17], the have-nots will be even more unable to afford sending their children to college, and may become still angrier at the establishment who they believe is ignoring them [18]. All this and more such ilk does not bode well!

Being an affluent nation does not automatically secure healthcare service quality. Organizing quality through use of IT need not be expensive if effectiveness is, as it should be, measured principally as customer (in this case, patient) value. This organizing effort, at a minimum, strategically requires proactive, purposeful, and high-minded networking between the operational healthcare providers and relevant context-forming government institutions.

3.2 Here Are Some Hopeful Suggestions

Quantum leaps in achieving better healthcare can only occur through behavioral change, and behaviors will only change with massive transformation of incentives. Most people will not pursue moral imperatives to operate beyond the current status quo of how their performance is measured and sustained without being rewarded somehow for doing so.

As has already been stated here and elsewhere there is a crying need to do better in sharing health record information. This can easily be done with current technology. What is more difficult is to break down institutional barriers that protect this information, that clearly took some investment to gather, from disclosure without some form of compensation. Hospitals should increase their degree of sharing while expecting those institutions they share with to reciprocate. If/when they do this mutual trust will build to further increase the "win-win situation" for the benefit of patients on both sides. There also must be an effort, which almost goes without saying, that certain personal information about the patients should certainly be withheld to protect their privacy. However, their basic medical data, focused on ailments that are useful for treating others, mainly from a database perspective, are what is critical. Naturally, considerable effort must be expended by the stakeholders to decide just which subset of information should be: (1) withheld, in general; and (2) which should be shared. And it could well be that another plethora of data could be in limbo until it is decided how to generate additional information, and into which of these two bins that subset should be placed.

Much can be done to reduce drug costs. Government subsidies have already been shown to help, particularly with the elderly and other disadvantaged patients. First and foremost, doctors should be more assertive in convincing patients to take the many lower-cost drugs that are just as effective as the more expensive drugs pharmaceutical companies are pushing in their advertising, especially on television (TV). In addition, many name-brand drugs are available in Canada, sometimes at ten percent of the cost of the same drugs in the US. Members of Congress must do better in exhibiting (hopefully, their innate) integrity tangibly in caring more about the health of their constituents than in bending to the temptations of lobbyist donor contributions. Representatives and Senators can be reelected in other ways than by outrageously expensive and offensive TV ads. For example, taking meaningful actions that help keep their constituents healthy can be explained in town hall meetings and the like. To the extent these actions are going in the proper direction, their constituents would likely spread this good news like wildfire via word of mouth and, indeed, via TV interviews once the media "steps up" to support such a trend!

Strengthening alliances among hospital administrators, doctors, nurses, patients, families, etc., can make the difference. Better healthcare practices start locally. It's all about more meaningful action-oriented communication and implementation of better ideas.

Does change, following the dominant deterministic model of management, really need to come from the top, and will this soon even be feasible? The second author happened to have an informal conversation with a local hospital administrator. He refuted our contention that costs would decrease if doctors would keep people healthy rather than just treating the sick. He countered, if someone does not receive healthcare and consequently dies of a heart attack, there are essentially no costs; whereas, if someone seeks consultations for a "bum" knee, say, subsequently undergoes remedial physical therapy, then a cortisone shot, yet needs an eventual knee replacement, that is quite costly.

As in complex systems, more significant experimentation with changes that may yield positive improvements will likely come from the bottom, especially considering the moral imperatives for keeping people healthier. Having governmental cutbacks on education and other social costs to pay for healthcare is not the way to go!

Again, behavior change is the key. More doctors are advocating healthy food as a medicinal remedy. Mandated societal changes can also be quite powerful in achieving heath gains. In Amsterdam, Copenhagen, and other European cities, for instance, there are thousands of bicyclists. Sure, there are a few bicyclists in the US but we are still quite wedded to cars and Sport Utility Vehicles (SUVs), and eschew public transportation.

The US spends comparatively much less on social services than several other Western countries. Thus, if one compares total healthcare costs more fairly, perhaps, the US actually spends less than approximately twice as much, as alluded to earlier. Also, again touching US politics, if healthcare costs are considered as a percentage of GNP, this statistic for the Affordable Care Act (ACA) ("Obamacare") has been essentially "flat" [14].

The prevailing trend of training more specialists than general practitioners (GPs) is starting to be reversed. Evidently, new doctors are now largely not going into fee-for-service [14]. Clearly, technology can be better applied to free-up doctors to spend more quality time with their patients, other than the few minutes (on the other side of waiting room, technician prep, and examination room waiting times) currently offered by many specialists.

And there are other significant efforts to move away from the more or less standard fee-for-service. For example, there is "bundling": "Bundled payments are a mechanism to deliver a single payment to providers for all the care an individual receives as treatment for a specific acute event or chronic condition. The benefits to health plans can be substantial, including: market incentives for providers to deliver efficient care and encourage patient compliance, reduced variation in the process and cost of care, and improved care coordination resulting in better quality" [19].

Capitation is another initiative that is being implemented: "Capitation is a fixed amount of money per patient per unit of time paid in advance to the physician for the delivery of health care services. The actual amount of money paid is determined by the ranges of services that are provided, the number of patients involved, and the period of time during which the services are provided" [20]. There are quite a few healthcare examples of doing better. Realistically though, broad, significant changes typically take about four years before they catch on, and ten more years are necessary for them to be more completely instituted. After all, it's a $3.4T industry [14]. Then there is the big idea of "single payer" which is an anathema to conservatives, generally, especially because "socialist" Bernie Sanders has been one of the strong advocates. If one accepts most of what conservatives would dismiss as untrue propaganda from organizations like Public Citizen [21], there is much that could be said logically and logistically, at least, about how there could be tremendous savings in current administrative and insurance costs by marginalizing the big insurance companies and providing Medicare for *all*, so the risk pool is appropriately widened to include mostly healthy people. Unfortunately, selling this to the young and healthy, who may have some difficulty contemplating future personal health problems, requires an appeal to what *could* happen to them as well as their moral sensibilities.

3.3 Connectivity, Big Data, and Information Technology

The most basic Internet of Things (IoT) concept is to enable and nurture *connectivity* among a multitude of autonomous electronic devices for human benefit; this: (1) could simplify our lives; (2) can be financially profitable; (3) offers pathways to professional satisfaction; and (4) presents cybersecurity opportunities. But is this all good? What about potential unintended consequences which may seriously *degrade* our quality of life? Will the IoT, even if implemented well in the face of formidable technical challenges, really improve our quality of life? We need to ask about and firmly establish the essential roles, cares, and desires of *people* in this world of automated artefacts.

This ethical concern can be elaborated briefly, for example, addressing each of the four motivations listed just above ([10] provides more detail.): (1) one automation-related outcome of IoT is that robots are taking over more and more jobs [17]; (2) many organizations care more about unlimited material growth [22] than improving the lot of humankind; (3) individual efforts often tend to divide people rather than bring us together; and (4) security is hampered by trying to protect too much instead of realizing the benefits of information sharing [1].

We are also critical of the BD term since the phenomenon it represents cloaks the true nature of the electronic information that should be embedded in many applications, particularly supply chains. In most services, it is not the vast bulk of hidden data, but real-time, *carefully selected* data of production value to agents and changing customer value perceptions, that need to be mined. Again, creating strategic connectivity should include elaborating on the role of people in the healthcare production system.

Knowledgeable healthcare stakeholders, after determining what works and what does not, should contact state-of-the-art technologists for help on what IT enablers might be tried. All this depends on sharing information, building trust, and striving for open mindsights, keeping the health and well-being of patients as the ultimate goal. This will more clearly define the system roles of value-producing agents.

An important and fundamental task in any supply chain is suitably integrating its diverse agents. One of the primary goals of recent years is improving the appropriate and worthwhile sharing of medical data. Doing so represents an expression of healthcare supply chain integration. Although IT is capable of this, as already stated, institutional profit barriers tend to hamper its achievement in healthcare services. And again, patients need to share at least their trend-able data to help improve general public health.

Safeguarding caregivers from routine mistakes in treatment is another critical area for improvement. IT should be able to revamp existing checklist processes in a simple, less time-consuming, and more accountable way. ISs of this sort should be designed to relieve doctors, nurses and technicians from time-wasting data-entry and retrieval tasks.

The target logistical IS must capture, store, and communicate to other agents, dynamically, data about resource availabilities and usages. So, when looking to apply the latest medical tool, IT, or IS as part of any healthcare endeavor, examine the pros and cons from this point of view. Important advances in this direction can be made considering very recent work that leverages cloud-based information sharing to improve hospital supply chain performance [23].

Exorbitant prices are imposed by drug companies and their lobbyists [24]. Again, as we have stated, patients need to largely ignore propaganda, about medicines they hear about on TV, for example, and doctors need to prescribe the much less expensive but effective alternative medicines.

In the provision of healthcare services, the questions of ethics should focus mainly on the well-being of the patients which includes, of course, lowering costs. This evidently includes eliminating, or at least mitigating, fee-for-service specialties. In terminal cases is not it better to provide services that induce feelings of comfort and contentment rather than more pain and suffering [25]?

Clearly, different incentives for all stakeholders are needed. Doctors should be rewarded for keeping people healthy. Hospitals should no longer be run as profit-making businesses. Insurance and pharmaceutical companies should operate in the public interest; politicians need insulation from lobbyists. We probably need to approach healthcare that people can afford, no matter where in the world they reside.

Additional guidance on these topics and managing supply chains can be found in [10].

4 SPECIFIC EXAMPLES

Reference [6] advocates a case study approach in learning more about complex systems and furthering complex systems engineering, and contains 23 case studies, one (Chapter 7) having much to do with healthcare. A case study template that can be used as a guideline in reporting on case studies of complex healthcare systems can be found in Chapter 3.

Reference [10] is focused on three (actually mini-) case studies central to the healthcare supply chain. In the next subsection, we summarize these as typical examples of what healthcare professionals face in their daily work lives. Together these cases provide different windows into issues regarding integration and information connectivity in healthcare service supply chains.

4.1 Case Study Overviews

The first case provides a narrative of commodity flow at one of the state-owned hospitals in Thailand. To understand the logistics flow of medical commodities in the hospital, one type of frequently-used medicine was selected and studied within its supply chain. A focus group, reviewing and assessing the results in depth, reached consensus on a promising way forward.

Managing material flows is a core feature of logistics. Here the flow of goods is a supporting but not core element, given that we are studying a services industry. One of the key aspects of this flow is the overstocking problem. One factor that contributes to the exacerbation of the problem are the natural and persistent time delays that may occur between steps in the overall logistical supply chain process. Possible mitigations to this phenomenon would be to endeavor to apply IT more effectively to not only reduce the manual labor of the participants but also to speed up the process, thereby reducing the human errors as well as the time delays.

The second case concerned waiting lists, specifically the treatment of a psychiatric patient focusing on consequences of unexpected and poorly informed changes of that type of flow. This involved an authentic incident that occurred in an outpatient psychiatric clinic in a Norwegian Hospital Trust, and represents customary priority setting procedures. Again, much was learned that benefited an approach to future improvements.

This case highlighted a problem of "falling through the cracks." A doctor scheduled to treat a patient got sick beforehand, an event that was unanticipated by the system. Thus, a seeming paradox ensued, a patient was recorded as finishing treatment in the general waiting list, despite still being ill, because the status of the patient recorded in a local waiting list was not taken into consideration as data in the general waiting list.

The third case addressed the extent to which effective communication exists in emergency situations, in particular, the standardized organization of information flow related to moving patients between hospitals and home care facilities in Norway. This took place when GPs and nurses in a home care system in a Norwegian municipality described how they communicate with the staff at a state-owned hospital. Opportunities for improving IT supported improvements were illuminated, for example.

The hospital physicians, GPs, and home care nurses usually communicate by electronic messaging. Yet these messages and even phone calls often used for clarification, do not accurately impart and capture all the relevant information and knowledge required. In addition, the nurses perceived an increased risk of misunderstanding if there were no inter-professional/personal contacts or meetings. Therefore, both nurses and physicians advocated for the importance of working together in person which naturally leads to further clarifications as well as new information and knowledge.

Reading more about these (mini-)cases and the issues they raise, consider how they could be expanded into full-fledged case studies.

4.2 Case Study Summary

There is a dearth of good ways to clarify responsibilities among the different professions involved. Communication is often cited as the most important factor for inter-professional cooperation, both by the nurses and even more by the GPs. However, electronic communication cannot substitute for the positive effects of meeting each other face-to-face to experience the often subtle but invaluable cues.

In a complex system, management must consider how to handle processes where components are continually in flux regarding not only the more easily measurable time, place, and form features, but also how they are interconnected (pooled) as well as how they are perceived. In this supply-chain, viewed as a complex system, logistics is an emergent process because the services and how they are evaluated over time are both transformed in each stage of production. Therefore, co-creation in healthcare logistics services is important. The role of IS here is to support process emergence. This involves data registration, informing each other about this emergent process's properties, and thereby supporting agency in sensemaking and managing this process in real time. Connectivity is a key attribute of such an IS, leveraging computer-human interfaces expressed as streamlined and visualization techniques that interconnect different supply chain agents. This means seeking and implementing an ethically viable potential for automating agency in the healthcare service supply chain where a large number of agents can interact in a non-deterministic way.

5 SUMMARY

Healthcare logistics is suggested as a complex system to operate and manage healthcare services in any economy regardless of its stage of development. This is particularly because human beings, viz., patients, physicians, nurses, suppliers, and administrators are centrally involved in providing and experiencing a variety of services and remedial treatments. Being predominately reciprocally interdependent, interaction in healthcare systems is reliant on intense exchanges to mutually adjust the relationships between patients and healthcare service providers. The pressing question regards the quality of this interaction, especially how long it takes to treat, and hopefully cure, the patient.

Existing policies and procedures are oriented mainly toward helping sick people to get well instead of instituting incentives and rewards for keeping people healthy as the higher priority. Admittedly, migrating healthcare more toward the latter mindsight is largely in the domain of government legislative and institutional executive actions and beyond the scope of logistical supply chain improvements. But healthcare providers already have the power to take a more holistic approach in establishing improved methodologies that can facilitate, along with carefully selected IS technologies, more effective interpersonal communication and networked interactions among all parties and elements of this complex system. That's what has been emphasized and explored here.

According to the literature and direct observations of existing practices, most healthcare services are well-intentioned and at least partially effective. However, it is also clear from interviews and analysis that systemic and systematic flaws significantly hinder higher quality healthcare, particularly in situations where time is a critical factor, e.g., in emergencies. Sure, one might prevent many emergencies by keeping people healthier but that is not the way the system operates, as we have outlined in Subsections 2.2 and 3.1. So, taking a broader and more humanistic view of the logistical supply problem, leveraging pooled as well as reciprocal interactions, emphasizing additional attention to contingencies, and better applying IT are among the advocated principles to help caregivers deliver better services, their primary function, while alleviating their secondary administrative burdens.

Since there is not yet a well-established and accepted theory that thoroughly informs healthcare service supply chains, case studies are of paramount importance in creating desirable outcomes. We really would like a better understanding of what works—and what doesn't—in practice, and good case studies can usefully illuminate many issues. Three subcases, the: (1) logistical flow of commodities at a hospital; (2) waiting list paradox; and (3) emergency flow and inter–professional communication about patients moving between hospital and home care, have been described and analyzed [10]. These should help set the stage for additional examples and more in-depth case studies in future work.

ACKNOWLEDGEMENTS

The case narratives provided in Section 4 are based on work-in-progress, where Ingunn Mundal and Trude F. Eines of Molde University College in Norway, and Duangpung Kritchanchai of Mahilol University in Thailand, provided the narrative based on their investigations. This work, as detailed in [10] is currently in the process of being published in an academic journal. The authors also thank Chad Walker of Analog Devices for alerting us to the cogent, stimulating, and entertaining Intelligence Squared video debate [14].

REFERENCES

[1] White, B.E., A Complex Adaptive Systems Engineering (CASE) Methodology: The Ten-Year Update, Presented at *IEEE Systems Conference*, Orlando, FL, 18–21 April 2016.

[2] Bar-Yam, Y., *Making Things Work: Solving Complex Problems in a Complex World*, 1st Edition. NECSI Knowledge Press, 2005.

[3] Kay, J.J., An Introduction to Systems Thinking. *The Ecosystem Approach: Complexity, Uncertainty, and Managing for Sustainability*, eds. D. Waltner-Toews, J.J. Kay & N-M.E. Lister, Columbia University Press: New York, pp. 3–13, 2008.

[4] Kay, J.J., Framing the Situation: Developing a System Description. *The Ecosystem Approach – Complexity, Uncertainty, and Managing for Sustainability*, eds. D. Waltner-Toews, J.J. Kay, & N-M.E. Lister, Columbia University Press: New York, pp. 16–34, 2008.

[5] Funtowicz, S. & Ravetz, J., Beyond Complex Systems: Emergent Complexity and Social Solidarity. *The Ecosystem Approach – Complexity, Uncertainty, and Managing for Sustainability*, eds. D. Waltner-Toews, J.J. Kay & N-M.E. Lister, Columbia University Press: New York, pp. 309–321, 2008.

[6] Gorod, A., White, B.E., Ireland, V., Gandhi, S.J. & Sauser, B.J., *Case Studies in System of Systems, Enterprise Systems, and Complex Systems Engineering*, CRC Press, Taylor & Francis Group: Boca Raton, FL, 2015.

[7] Norman, D.O. & Kuras, M.L., Engineering Complex Systems. *Complex Engineered Systems: Science Meets Technology*, eds. D. Braha, A. Minai, Y. Bar-Yam, Springer: New England Complex Systems Institute, Cambridge, MA, 2006.

[8] White, B.E., On leadership in the complex adaptive systems engineering of enterprise transformation. *Journal of Enterprise Transformation,* **5**(3), pp. 192–217, 2015. http://dx.doi.org/10.1080/19488289.2015.1056450

[9] Gilbertson, R., Tanju, B. & Eveleigh, T.J., A complexity based heuristic decision analysis model to recommend systems engineering domain. *IEEE Engineering Management Review*, **45**(3), pp. 64–81, 2017. https://doi.org/10.1109/emr.2017.2734358

[10] Engleseth, P., White, B.E., Mundal, I., Eines, T.F. & Kritchanchai, D., Healthcare logistics as a complex system. *Production Planning & Control,* In review.

[11] Hawn, G. (with Holden, W.), *10 Mindful Minutes*, Penguin Books: New York, p. xiv, 2011.

[12] Ashby, W.R., Requisite variety and implications for control of complex systems. *Cybernetica*, **1**(2), pp. 83–99, 1958.

[13] Rzevski, G. & Skobelev, P., *Managing Complexity*, WIT Press: Southhampton, UK, 2014.

[14] Proposition, The U.S. Health Care System is Terminally Broken, *Intelligence Squared Debates*, available at https://www.intelligencesquaredus.org/debates/us-health-care-system-terminally-broken (accessed 21 October 2017)

[15] A Robust Health Data Infrastructure, prepared for: *Agency for Healthcare Research and Quality*, 540 Gaither Road Rockville, MD, 2014, available at www.ahrq.gov (Contract No. JSR-13–700, Prepared by: JASON, The MITRE Corporation, 7515 Colshire Drive McLean, VA 22102-750, AHRQ Publication No. 14-0041-EF, April 2014.)

[16] Epipen, 2017, available at: https://www.reuters.com/article/us-health-epipen-costs/another-look-at-the-surge-in-epipen-costs-idUSKBN16Y24O

[17] White, B.E., Fathoming the future of artificially intelligent robots. *International Journal of Design & Nature and Ecodynamics*, **13**(1), pp. 1–15, 2018. https://doi.org/10.2495/dne-v13-n1-1-15

[18] Bageant, J.L., *Deer Hunting with Jesus*. Three Rivers Press: New York, 2007.

[19] Bundling, 2017, available at https://truvenhealth.com/portals/0/assets/HP_11512_0812_BundledPayments_SS_WEB.pdf

[20] Capitation, 2017, available at https://www.acponline.org/about-acp/about-internal-medicine/career-paths/residency-career-counseling/guidance/understanding-capitation

[21] Public Citizen, mass mailing material, 1600 20th Street, NW, Washington, D.C., 2017, available at www.citizen.org

[22] White, B.E., Let's Do Better in Limiting Material Growth to Conserve Our Earth's Resources, *Conference on Systems Engineering Research* (CSER) 2012, St. Louis, MO, 19–22 March 2012.

[23] Kochan, C.G., Nowicki, D.R., Sauser, B. & Randall, W.S., Impact of cloud-based information sharing on hospital supply chain performance: a system dynamics framework. *International Journal of Production Economics*, **195**, pp. 168–185, 2017. https://doi.org/10.1016/j.ijpe.2017.10.008

[24] Abramson, J., *Overdosed America: The Broken Promise of American Medicine,* 3rd edn, Harper-Collins: New York, 2004.

[25] Gawande, A., *Being Mortal: Medicine and What Matters in the End*, Henry Holt: New York, 2004.

THE BEAUTY OF ARCHITECTURAL COMPLEXITY

ROBERT BARELKOWSKI
West-Pomeranian University of Technology, Szczecin.

ABSTRACT
The paper is an attempt to revisit the concept of beauty which is analyzed in reference to Vitruvian principles of architecture supplemented with idea, the fourth principle. The holistic approach to architecture may be the vehicle to explain universal patterns of beauty, possible to be understood as complex system – both the discipline and constructs created as a result of architectural ideas. One can argue that architectural entities start their existence with initial idea, then its purpose is defined, to be encased in structure and form. All these principal bricks architecture is built with exist on multiple layers, as abstracts or as physical manifestations. The beauty is therefore the perceptive phenomenon, in which man is able to decipher and comprehend expressed or tacit properties of the universe and convert them into his own interpretation of the environment and man-made culture or social patterns.

This paper is a tribute and humble contribution to pay respect to Professor Carlos Brebbia (1938–2018), whose grasp of complex reality opened many inspiring paths within science and animated multiple connections within scientific community for the benefit of knowledge and understanding of the nature.
Keywords: architectural idea (concept), architecture, beauty, complexity,

1 INTRODUCTION

The idea of beauty has always been at the center of theoretical considerations of what architecture constitutes. It contributes significantly to our understanding of the role of architecture, its impact or mutual connection between people and their creations, in this particular instance creations of a scale enabling them to alter the environment. Architecture appears to be a discipline of science and art at the same time, it provides common ground for antinomies – the domain of the objective, and the domain of the subjective. It blurs the boundaries and becomes an example of applied science which defies some of scientific principles, e.g. usual rejection of single-case-based theories or case studies which is hardly applicable in architecture – in case of this particular discipline most cases are unique and the process of extracting general principles, theorems or rules becomes vulnerable to approximations, analogies, uncertainties in causal connections.

When Gordon Graham ponders architecture as art, he rightfully sees architectural character as distinctive in comparison with other arts. Architecture raises several doubts about the applicability of aesthetical values as primary ones, because it is useful, purposeful, relevant, and it is both designed and built with the extensive use of ready-made objects [1]. Its usefulness is generally coerced, an inherent trait without which architecture loses its special status and may become something else, a sculpture, an object in space devoid of common meanings. This difference is at the same time significant yet intangible, without clear edges to what extent functionality defines perception of order and beauty. Functionality, being on the one hand related to intended purpose, and on the other hand attributed as a result of multiple unpredictable occurrences, exposes the inherent evolutionary process of how architecture matures, achieves its independence from creator, designer, in fact anyone involved in its inception into built environment. I would argue that what Nicolaus Pevsner defined as architectural "honesty", architectural truth, reflection of zeitgeist [2] is in fact the projection of

DOI: 10.2495/DNE-V13-N3-250-259

this particular feature the architecture has – the beauty within the innate order. It goes far beyond aesthetic properties, attempts to attach additional value to ordinary objects, enriches their values.

In this work the concept of architecture is investigated in context of its main components, which architectural theory in the past formulated as usefulness, durability, and beauty, supplemented with abstract notion of idea responsible for initiating and manifesting architecture in multiple aspects. It departs from tri-fold *utilitas*, *firmitas* and *venustas* and revisits modernist transformation of the three concepts. I argue that the notion of beauty implemented in architecture is different from the notion of form and aesthetics of architecture, and that even Vitruvian *venustas* was seen by the ancient author as indivisible from other values and in fact all them constitute the beauty and its definition as proposed herein. Therefore the paper starts with the introductory thoughts and perspective on this issue and provides considerations related to four architectural principles. There is superficiality and hidden qualities of architecture acting in unity, and their atomization, separation leads astray. The paper reviews four principles, discusses their interpretation, attempts to integrate these four elements, and proposes the insight into architecture relying on holistic view and exploits the perspective of complexity.

Architecture in its subtlety exposes human interpretation of holistic view of reality, including its holistic interpretation. This so-called truth, which Roger Scruton calls objectivity of architectural aesthetics [3], unveils the most fundamental significance of architecture, a discipline seemingly devoid of its own original, exclusive disciplinary field, one that is a combination of engineering, crafts, arts, rather than something autonomous. It is only an apparent lack of autonomy – its specificity lies in all-encompassing nature, in pursuit of synergy in preparation and performance of architecture. Thus the assertion that the beauty of architecture is in fact the reflection of its inherent complexity reveals the source of theoretical dilemmas in which beauty was mistakenly seen more as aesthetic attribute than multi-layered unity of various properties. This phenomenon, although diagnosed in different manner and described in different words, is a prevalent observation of many researchers of the theory of architecture, seen in the past as well as in the present [4].

One may argue that the concept of architectural beauty is more similar to ideas found in science, to the concept of the same beauty that is found in nature – hidden rules allowing for simultaneous existence of order and chaos supplementing each other and making nothing but superficial dichotomy discovered in the environment, regardless of whether natural or artificial (even virtual). The beauty found in scientific considerations, in mathematics or physics, as found in Heisenberg's memories [5], or the beauty discovered in sophisticated geometries of Parthenon in Athens which import distortions and deformations (related to human perception) necessary to transform perfect geometry into perfect beauty [6].

2 THE BEAUTY OF COMPLEXITY OF ARCHITECTURAL IDEA

The famous Vitruvian triad determines the most significant values of architecture considered throughout history of civilization. *Utilitas, firmitas et venustas* – the three principles that stem from human understanding of the environment, both natural and one being the result of man-made artifacts and entities. These principles offered short but accurate description of the nature of architecture – connection between inanimate objects and space imbued with purposeful thoughts and actions of people. It is important to see the intellectual domain as an integral part of the environment, despite the fact that it was obscured by modern revolution in the 20th century. This transformative perspective is necessary to explain the erosion of holistic interpretation of architecture so prevalent in classic, ancient times, and hence the absence of complexity issues until they modestly reappear at the turn of the 21st century with works by Christopher Alexander [7] and those by Michael Mehaffy with Nikos Salingaros [8].

It is significant because architecture is, obviously, connected to architectural theory and its philosophical background, it is immersed in multi-science domain, making the meeting point for sometimes seemingly contradictory aspects of real life as well as scientific viewpoints. Therefore, the deformation of architectural principles affects the interpretation of the theory and is reflected in simplification of perception of architecture, in the end leading to misunderstanding and misdirecting contemporary thinking. *Utilitas*, *firmitas*, and *venustas* have been replaced with new ideas of function, structure, and form. While it may seem innocent, it replaces fundamental and abstract values with their restricted, fragmentary substitutes. Function is not an equivalent of usefulness – it is its contemporary, thus temporary, manifestation. Structure doesn't equal *firmitas*, because structure is only appearance, an effect of complex process of architectural embodiment. Finally, the term "form" is not compatible with *venustas*, even apart from etymological roots of "beauty" reflecting the feature of an object being saturated with particular universal value.

It is the understanding of unity of things exemplified in Stuart Burgess analysis of the beauty of peacock's tail [9]. The aesthetic complexity is the result of multiple coordinated phenomena: the shape, material, reactiveness to light, beholder's eye structure, light transmission, bird's movement and behavior. The perception is reduced to the admiration of shapes, variety of glittering colors, namely optical effects produced by thin-film interference, but comprehension of this phenomenon is impossible without seeing the emergence of the beauty from joint impact of multiple factors, some of them physical, others ideological, derivatives from human contemplative approach to reality. The conceptual pattern hidden in reality is exposed by Murray Gell-Mann, who points out that the organization of the universe is recognized in the balance between regularities and irregularities in information chains coded in living species and inanimate things [10]. Gell-Mann finds this self-regulating balance in the organization of complex adaptive systems. While he points towards the ability to process regularities and produce (or re-produce) schemes prolonging either life itself, or its modes of organization, what is interesting is innate imperfection of the system, which is required to allow for change, for adaptation, for improvement and re-adjustment to changing conditions. Architecture, being human creation, mirrors these schemes, and also mirrors the innate imperfections which at the same time bless architecture with ability to thrive, transform, adapt, as natural systems do.

Dealing with the idea, with the information, is what defines the start of emergence of architecture – from the intent to the embodiment. Necessity to organize patterns of social life is reflected in architecture in a selective way, however still conforming to the principles of complex adaptive systems. It requires, however, both regularities converted into schemes and imperfections retaining the system as an open one, indeterministic, self-organizing. While the former is often apparent and clearly recognized, e.g. patterns recognized by Alexander's team [11], the latter is rarely seen as inherent or positive force of architecture. Here it is important to make distinction between, what Alexander describes as "dead" or "bad" patterns, and irregular patterns which are test platforms for change and in-programmed instability, because only these atypical schemes are responsible for vitality and validity of the system (of architecture).

The beauty of the architectural idea lies in complexity of the concept mirroring what the idea attempts to define, solve, what it strives to anticipate – and its aesthetics is an integral component of thinking, also outside of aesthetic criteria of designed object. It is more the beauty of integrity, unity, than beauty of appearance. Appearance is only a robe, a skin that tells us how profound this complexity is and how it affects external, formal aspect of architecture, how it build "architectural experience". As we know, this experience is not constrained to superficial reception of building or space – its anchored in constant perceptive mechanisms, in continuous life experience. This is well explained by Helmut Leder et al. in their model of aesthetic appreciation and judgment, in which they emphasize context and input for experience, perpetuality in performing analyses, implicit memory integration, explicit classification, evaluation, and affective / emotional processing [12]. Subsequent steps of this experiencing may be determined as, respectively, external, objective or semi-objective/independent factors corresponding to context and input, unceasing process (perpetuality), cumulative comprehension of reality, resultant from integrating memories, and often comparative processes which can be linked to classification and evaluation. In other words, we simultaneously perceive and assess, and both integrated activities are producing our comprehension of the nature of things – the evaluation of their beauty. The creative process, e.g. architectural designing, is simply mirroring appreciation and judgment – it also uses spatial and temporal context and multiple sources of input, it requires constant re-evaluation due to wicked nature of architectural problems [13], it enacts inner memory, social memory, and semiologic code (which could be associated with sedimentation of social memories inscribed in abstracted signs) [14]. This wickedness, these irregularities are generators of the random, nodes of emergence, loci of opportunities which sometimes lead to errors, but sometimes allow for creation of a masterpiece. The imperfection assures this evolutionary bias, reflecting our evolutionary thinking also in perceptive and aesthetic spheres of our connection to reality. As Paul Hekkert writes, patterns generated by humans have this innate trait of adaptivity [15], and hence any architectural idea also contains elements of unpredictability, random results, with perpetual adaptivity.

3 THE BEAUTY IN ARCHITECTURAL FUNCTION

Function is the aspect of architecture that is immediately associated with usefulness, with practical performance of space. It is expected to provide the content, the result of transposition of an idea into something working, something practically responding to requirements. Function represents the role of space of becoming socially driven vessel, containment for different processes and interactions for which function-related labels are merely signaling the purpose – to give an example of such performance one may still refer to what Graham wrote on the purpose of architecture. He argues rightfully that while there is an initial purpose in the inception of piece of architecture, it is not permanent, it changes. At the same time he points out that future change of purpose or even seeming lack of thereof does not change this core value of architecture attributed to inherent potential of having a purpose [16]. Contrary to Graham and other researchers like Scruton, however, I would like to propose viewing particular sites as more than just arrangements in space conceived for admiration as they claim in case of those non-architectural places. Particularly I insist on rejection on Karstin Harries simplified interpretation of decorated shed, architecture reduced to the role of container [17]. Even if this is eligible from the perspective of designer allowing for specific approach to architecture as boundary, permissible for out- and inflows, restrained to determination of how transition between outside and inside grants the architect a comfort of dealing with limited number of problems, similarly to temporal aspect of architecture it doesn't refute unity of shell and content.

Function embodies specific architectural code inscribed in building to fulfill the tasks defined by the idea. It is not the set of volumes strung on some lines representing use-related patterns. These are more meaningful chambers conveying different messages, in good cases coherent with the entity, in bad ones dissonant. Herein evaluation appears, our perception of "appropriateness" of the function, connected to the appreciation of beauty of functional structure. Function is strong reference to beauty of the object, it initiates assessment and is inseparable from aesthetic principles, although not necessarily must play primary role [18]. Function brings understanding of internal structure, it brings comprehension of unifying qualities which make purposeful chains valid and beautiful. There are simple questions illustrating this unity – can we ignore aesthetic component in how function is distributed within architecture? Can we avoid calculating the importance of aesthetics of linking volumes in particular sets or chains to facilitate specific uses? Aren't these chains manifestations of inner order, profoundly connected to the beauty of spatial arrangements?

If we are unable to find answers instantly, the unstable elements of functional program of architecture give us the clue about the relevance of beauty. While idea quickly becomes something blurred or even fades, function remains, but fluctuates or at least is prone to fluctuations. This openness of functional program is unable to include all implicit possibilities of use, yet designed space purposefully or intuitively provide these possibilities, makes architectural object to deny any attempt to achieve equilibrium and promotes contained or forced self-organization. Alterations in program explore ability of architecture to absorb and use specific connection between program and volumes. While idea is responsible for insertion of intended elements as well as, function exhibits human-related order of architecture, the order of human interactions and processes.

4 THE BEAUTY IN COMPLEX STRUCTURE

Structure is different from function (and idea) significantly in that, apart from the processes of decay and decomposition, it is usually fixed and its potential capacity most often does not involve reconfiguration. The backbone for architectural form, structures do not divert from developing the aesthetically loaded constructs. If natural beauty is taken into account, it is next to impossible to point out any element which would be stripped from aesthetic content, although this content can be interpreted within the autonomy of any part, within the connection between part and entity, as well as within entity perceived in its entire spectrum of existence.

Interestingly, structural thread of these considerations reveal relationship between physical construction and (abstract) problem of truth. The architectural structure must follow rules of logic and support the performance – distribution of atoms support the rigidity of a crystal, leg bones assure dexterity and appropriate movement of a cat, structural veins of moth's wing granting lightness, flexibility, and extremely good aerodynamics. The elements of construction in architectural object play similar role and also express the purpose, the logic, the justification for existence. This kind of rapture becomes integrated on the field of bionics and bionic design, implying its Latin meaning – the way of life. In this way truth of life is revealed, natural laws get exposed, and examination of natural world and purposeful design of sophisticated structure for architectural building have this urge to understand universal principles and the science of the world in common. While idea is inherently complicated and function has little possibility to avoid similar trait, while form may be independent, structure is a component prone to optimization [19]. The nature, being the reference for humankind, shows how to compress, to use as little material as possible, how to innovate with materials [20]. Lan Ma reviews different levels and layers of implementation of bionics, and determines also its structural importance [21], and what can be seen is intuitively apprehended image of fascinating energy distribution of the system. Guy Battle and Christopher McCarthy describe it as sculpting with energy [22]. It may be easily observed in Antonio Gaudi's work for la Sagrada Familia, the hanging chain model of gravity lines turned upside down to expose the optimal energy distribution of the system necessary to be converted into physical, loadbearing matter [23].

So is there a beauty of structure? And if so is beauty an aesthetic concept or holistic concept in which one cannot separate the aesthetic component from the purpose and the performance of the object? Architectural structure may not be seen, but regardless of its visibility it is intuitively assessed, examined in terms of truth versus lie or disguise. Contemporary architecture, using more organic, sculptural and also total forms in its appearance is, to some extent, a reflection of natural world, an imitation, sometimes a simulacrum, purposefully transformed and adjusted. And the language of new forms is in fact a closer approximation of the nature, the import of natural complexity integrated in shape – one which contains function, provides shell for the content, supports itself. Structure doesn't change or change a little, but this doesn't disprove its share in complexity – the compactness, thinning down structural volume only ever allows for acquiring sophisticated, yet purposeful architectural shells. This co-evolutionary trait and connectivity linking fixed or mobile but sustaining structure to other components probably was never sufficiently explored. Structure adds logic of physical, material organization as reflection of human-related organization of spaces. Function is rightfully seen as "negative" of structure – matter separating or joining contained processes. Thus it becomes clear that instead of dichotomy these aspects explain the same idea, one following the other and *vice versa* – without any precedence.

5 COMPLEXITY OF ARCHITECTURAL FORM

It seems that understanding of complexity of form is the most apparent, simple task in comparison to other components of architecture. However, at the same time, the notion of form is burdened with aesthetic content apparently replacing and erasing any competition. Form, if not restricted to shape resultant from construction, is seen as decorative, serving the purpose of admiration. This narrow view can be easily refuted by parallel to natural world, because after all social groups have some behavioral similarities to what can be observed in nature – the sense of place, urge to congregate and socialize. It's attractive abilities can remind of tree perfect for providing observation point for birds who appear there in flock (let's treat it as metaphor, not a literal association, with all provisions given by Bruno Latour while discussing social interactions and distinctive nature of human experiential and cognitive abilities in such interactions [24]).

While many people interested in architecture may perceive that its output is brave and avant-garde, and that architectural trends are set by designers and their inventive, creative inner-selves, it is the opposite. The majority of architectural forms, even if produced by world renown brands, generate functional buildings, and just few are planned to be architectural icons – through careful process of economic planning, site selection, budget construction etc. Special conditions must be met to allow for such a result, and then the role of an attractor, a gathering point, a place to be in, to engage socially, is the major factor justifying any fanciful attributes architecture can have formally. So this apparent fancifulness is a result of complex interconnected social, economic and spatial processes usually far from being whimsical, even if we acknowledge some failures or exaggerated spendings. Processes are embodied in form which reflects all these phenomena in a sophisticated, complex manner, filtered through designer's interpretative imagination/thinking, transposed into structure and shell intended to express the meaning, the content, and the promise of undetermined occurrences. Fortunately not all criteria must be reflected in equally masterful way and there is no need for any standard or dogma, but still holistic approach to beauty, called by Mike Schlaich "elegance" [25], is required. And architectural form triggers or responds to some social behaviors, accommodates them (by its attractiveness) or even creates. It is much more important for a form to have an implicit meaning, influence, and associative powers, than to express shape, color, and material, although the latter are necessary to convey the former.

The architectural form is addressed to everyone ready to receive signal, message, and to process it, interpret, evaluate – it is not admiration *per se*, it is instant comprehension of inner complexity of form as a reflection of interwoven threads of reality, again.

6 PROFOUND PERCEPTION OF COMPLEXITY?

We have reviewed different aspects of beauty of architecture and architectural entities. Vitruvian triad has been extended to initial component of architecture – the idea (ideological content). The inclusion of an idea is herein postulated as compulsory, because the immaterial is equally important in architecture and influential for its existence and performance as remaining, traditional virtues discussed in the ancient times by Vitruvius. The code to decipher this beauty is complexity, or maybe it would be more precise to say that beauty is what emerges from perceiving the code of complexity. Architecture doesn't mimic socio-spatial reality, it interprets it and exposes coherent (or incoherent) vision of social organization and uses tools defined as principles of complexity [26].

Architecture emerges from idea through purpose (function), and then it is nested within structure and manifested by form – forming link between immaterial and material, psychological and physical, mind and body, completed with subjective, personal experience acquired through reception of integrated entity. The user, the recipient, the observer receives instant, compacted information containing traits of all these four components. The perception of beauty becomes a discovery of truth hidden within social and cultural patterns, of how they work, how they interact and how they are composed into systems or sub-systems, analogous to complexity of universal, natural clusters or systems. This phenomenon touches human beings as well as constructs or other objects. Emilio Del Giudice and Alberto Tedeschi notice that human beings, as other living organisms, are coherent systems open to the external world and consequently be able to tune their own oscillations with the fluctuations of some part of the Universe [27] – and the feeling of beauty may be therefore interpreted as the decoding of this connection. Architecture becomes a vessel for anthropic perspective of understanding the sense of being, the sense of coherency present in all-encompassing, unified reality – all layers of this reality recognized with transdisciplinary approach required to understand holistic image: it builds the ontological axiom, the logical axiom, and the complexity axiom, as Basarab Nicolescu puts it [28]. The beauty is the knowledge of interdependencies (having similar properties to scientific knowledge, inherently contaminated with subjective perspective of a researcher – there is no objective science, only objectivized, although we strive to achieve this objectivity), simultaneously internal and external [29].

We observe in architecture (and in architectural objects) what Michael Baranger writes about complexity: non-linear interaction of constituents (or components), their interdependency, multi-scalar structure, generation of actions or behaviors, integrated connection or unity of chaos and order (which Baranger calls non-chaos), coexistence of competition and cooperation – integrated reinforcing or weakening impact of architectural factors [30]. And when Stanley Stein and Michael McMordie critically assess the loss of central role of beauty, they imply not the superficial concept of beauty, rejecting the formalistic approach to this notion. They quote an important thought once put forth by Ludwig Wittgenstein referring to common attributes sought for in human rapture with reality that it is similarities and relationships [31], not repetitive, common patterns that build multidimensional experience of beauty – people communicating their understanding of nature enriched by social and cultural patterns through architecture, to each other and with the purpose to mark their presence. The imprint of primal, natural imperative can overcome limitations of assessment and knowledge as phenomena attached to particular spatio-temporal reality, thus to particular moment and particular being – because they have universal framework.

It is hard to believe that this approach to the notion of beauty can conclude or narrow down the discussion on how beauty should be interpreted. However, too many interpretations are fragmentary, focused on limited concept of what architecture is, particularly when it is perceived as formal expression of inanimate matter in which proportions or even mathematical relationships are treated instrumentally without looking for their hidden meaning. I would argue that architecture which is the unwanted muse among others, those who are representing seemingly more autonomous arts, requires more profound perception, and that this perception can be achieved through complexity and its principles. It seems that this framework and multi-layered picture of complexity and its image – the experience of universal beauty – can, to some extent, help overcome multiple apparent contradictions within the theory of architecture.

REFERENCES

[1] Graham, G., *Philosophy of the Arts: An Introduction to Aesthetics*, Routledge, Oxon and New York, 2005. See pp. 165–168.

[2] Pevsner, N., *The Cities of London and Westminster: The Buildings of England*, Vol. 1, Harmondsworth, Penguin, London, 1972.

[3] Scruton, R., *The Aesthetics of Architecture*, Methuen and Co.: London, 1979. See pp. 263.

[4] Hendrix, J., Theorizing a contradiction between form and function in architecture, *South African Journal of Art History*, **27**, pp. 9–28, 2012. (Cf. discussion on Quatremere de Quincy's views on classical Greek architecture, which, in his eyes, integrated the domain of the physical and the ideal. See p. 12.)

[5] Heisenberg, W., *Inner Exile*, Birkhauser, Boston, 1984. See pp. 143–144.

[6] Tatarkiewicz, W., *Historia estetyki. Estetyka starozytna*, Vol. 1, Arkady, Warszawa, 1985. See pp. 57–59.

[7] Alexander, C., New Concepts in Complexity Theory Arising from Studies in Architecture: An Overview of the Four Books of The Nature of Order with Emphasis on the Scientific Problems which Are Raised, *Katarxis*, No. 3, 2003, available at www.katarxis3.com/ (accessed 10 March 2018, pp. 24. See 7–8.)

[8] Mehaffy, M. & Salingaros, N.A., Architectural myopia: designing for industry, not people. *On the Commons*, pp. 5 & 9–10, 2011.

[9] Burgess, S. C., The physical structure, optical mechanics and aesthetics of peacock tail feathers. *Design and Nature, WIT Transactions on Ecology and the Environment*, Vol. 57, eds. C.A. Brebbia, L. Sucharov & P. Pascola, WIT Press, Southampton and Boston, pp. 435–443, 2002.

[10] Gell-Mann, M., Plectics: The study of simplicity and complexity. *Europhysics News*, **33(1)**, pp. 17–20.
https://doi.org/10.1051/epn:2002105

[11] Alexander, C., *The Timeless Way of Building*, Oxford University Press, New York, 1979. See pp. 126–136.

[12] Leder, H., Belke, B., Oeberst, A. & Augustin, D., A model of aesthetic appreciation and aesthetic judgements, *British Journal of Psychology*, **95**, 489–508, 2004.
https://doi.org/10.1348/0007126042369811

[13] Webber, M.W., Planning Problems are Wicked Problems. *Developments in Design Methodology*, ed. N. Cross, John Wiley and Sons, Chichester, pp. 135–144, 1984.

[14] Barelkowski, R., Verbal thinking in the design process. *Design Principles and Practices: An International Journal*, **4(5)**, pp. 127–138.
https://doi.org/10.18848/1833-1874/cgp/v04i05/37953

[15] Hekkert, P., Design aesthetics principles of pleasure in design. *Psychology Science*, **48**(2), pp. 157–172, 2006.

[16] Op. cit.: Graham (2005: 167).

[17] Harries, K., *The Ethical Function of Architecture*, MIT Press, Cambridge, 1997.

[18] Sauchelli, A., Functional beauty, architecture, and morality: a beautiful Konzentrazionslager? *The Philosophical Quarterly*, **62(246)**, pp. 128–147, 2012.
https://doi.org/10.1111/j.1467-9213.2011.00020.x

[19] Sadri, M., Kavandi, M., Jozepiri, A., Teimouri, S. & Abbasi, F., Bionic architecture, forms and constructions. *Research Journal of Recent Sciences*, **3(3)**, pp. 93–98, 2014.

[20] Battle, G. & McCarthy, C., *Sustainable Ecosystems and the Built Environment*, Wiley-Academy, Chichester, 2001. See pp. 15–17.

[21] Ma, L., From Bionic Design of the High-Rise Buildings, *2nd International Conference on Civil, Materials and Environmental Sciences* (CMES 2015), London, the UK, pp. 730–734, 2015.

[22] Op. cit., Battle and McCarthy (2001: 102–103).

[23] Huerta, S., Structural design in the work of Gaudi. *Architectural Science Review*, **49**(4), pp. 324–339, 2006. See pp. 330–332.
https://doi.org/10.3763/asre.2006.4943

[24] Latour, B., On interobjectivity. *Mind, Culture, and Activity*, **3(4)**, pp. 228–245, 1996. See pp. 229–230 and 239–240.
https://doi.org/10.1207/s15327884mca0304_2

[25] Schlaich, M., Elegant structures. *The Structural Engineer*, 2015, pp. 10–13. See p. 12.

[26] Barelkowski, R., Design process as complex system. *International Journal of Design & Nature and Ecodynamics*, **13**(1), pp. 46–59, 2018. See pp. 48–51.
https://doi.org/10.2495/dne-v13-n1-46-59

[27] Del Giudice, E. & Tedeschi, A., Science and aesthetics: how beauty can emerge from matter. *International Journal of Design & Nature and Ecodynamics*, **6(4)**, pp. 282–285, 2011. See p. 283.
https://doi.org/10.2495/dne-v6-n4-282-285

[28] Nicolescu, B., Methodology of transdisciplinarity – levels of reality, logic of the included middle and complexity. *Transdisciplinary Journal of Engineering & Science*, **1(1)**, pp. 19–38, 2010. See p. 24.

[29] Ibid. 27.

[30] Baranger, M., *Chaos, Complexity, and Entropy: A physics talk for non-physicists,* 2000, available at necsi.edu/projects/baranger/cce.pdf (accessed 3 April 2018, pp. 735–747. See pp. 741–742.)

[31] Stein, S. & McMordie, M., Beauty and character: values beyond beauty in architecture. *Interchange*, **33(2)**, pp. 183–191, 2002. See pp. 184–185. https://doi.org/10.1023/a:1016552320002

DIGITAL TWINS FOR DECISION MAKING IN COMPLEX PRODUCTION AND LOGISTIC ENTERPRISES

WOLFGANG KUEHN
University of Wuppertal, Wuppertal, Germany.

ABSTRACT

The operation and optimization of complex production and logistic enterprises requires modern, data and simulation driven multi criteria decision approaches. Digital enterprise systems are complex systems, in terms of layout variability, control strategies, business processes and system parameter. All of these aspects are not independent and even due to the system dynamics the optimal solution may differ depending on the actual situation applying specific requirements. In order to improve operative decision making in production and logistic enterprises a digital twin concept can be applied. These digital twins are virtual clones of real systems or subsystems. With simulation and artificial intelligence technologies new opportunities for operative decision making are possible. Digital twins can cover the entire lifecycle of an asset or process by forming a closed-loop chain for smart, connected products, services and production and logistic processes, from design to operation, from deployment to continuous improvement.

Keywords: Digital Enterprise, Digital Twin, Factory Simulation, Virtual Model

1 INTRODUCTION

Digitalization changes everything rapidly, it changes the daily lives as well as existing business models and creates new business opportunities. Products from all industries are getting more innovative, more data centric, and involve more software than ever. With the Internet of Things (IoT) [1] the connectivity becomes an integral component and the next generation of smart products will be more complex systems that make current development processes inadequate [2]. The industrial environment is becoming more connected. Intelligent devices and machines generate massive amounts of data, creating a bridge between real and virtual worlds [3]. Turning data into value is getting a key success factor.

2 COMPLEX DIGITAL ENTERPRISES

Modern digital enterprises require a holistic approach to optimize the entire value chain [4]. By use of integrated software solutions a digital image of the entire value chains can be created [5]. These digital enterprise suites have to offer an integrated portfolio of software-based systems and automation technologies for industries to seamlessly integrate and digitalize the entire value chain, including suppliers. With a Digital Twin concept, a virtual copy of the value chain can be crated [6]. This digital clone allows to simulate, to test, and to optimize in a completely virtual environment in order to reduced time to market, increased flexibility, quality and efficiency. The Digital Twin concept enables optimization in the virtual world and the subsequent transfer of results to the real world.

The operation and optimization require modern, data and simulation driven multi criteria decision approaches. Digital enterprise systems are complex systems, in terms of layout variability, complex control strategies, business processes and system parameter [8]. All of these aspects are not independent and even due to the system dynamics the optimal solution may differ depending on the actual situation and specific requirements.

DOI: 10.2495/DNE-V13-N3-260-271

Figure 1: Enterprise Requirements (*Source:* Siemens 2017 [9])

2.1 Enterprise Requirements

In digital enterprises the engineering process has to achieve competitive advantage through improved engineering performance. Innovative engineering accelerates the design of the product and production process by enabling team collaboration to streamline the engineering processes.

In enterprises the demand for improved product performance, production flexibility and reduced operating cost are driving force of innovation, facilitated by collaborative program management across the product lifecycle and the supply chain throughout the entire product lifecycle. Within the digital enterprise the following areas are required:

- Product Lifecycle Management (PLM)
- Product Data Management (PDM)
- Manufacturing Operations Management (MOM)
- Electronic Design Automation (EDA)
- Totally Integrated Automation (TIA)
- Asset Performance Management (APM)
- Application Lifecycle Management (ALM)

Enterprises have to respond effectively to changing production schedules, which are made more challenging due to the requirements of global supply chains. The challenge of adopting new and innovative technologies and processes in order to compete successfully with increased efficiency and flexibility.

2.2 Enterprise Decision Levels

Complex enterprises require decisions regarding design, engineering, planning, communication and controls on different enterprise levels. Each of the particular decision requires specific algorithms and data. Therefore, standardized interfaces between these levels are mandatory.

Digital enterprises should use common data for all applications on different modelling levels in order to enable collaboration with virtual models for different purposes and different levels of detail [10]. An open architecture, open interfaces and interoperability are key factors for implementing digital enterprise concepts.

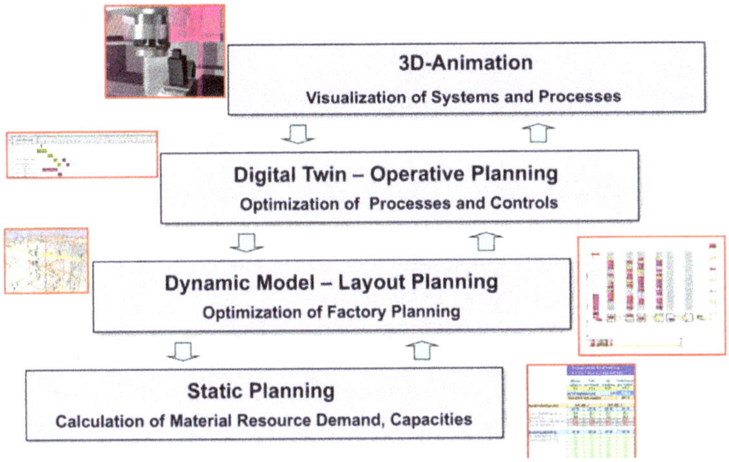

Figure 2: Enterprise decision levels

2.3 Enterprise Optimization

In digital enterprise systems the layout of manufacturing and logistics, the control strategies and a huge variety of parameters have to be considered and optimized [11]. Due to the complexity of digital enterprises the optimization of these systems requires multi criteria decision approaches. Most of the parameters involved are not independent and even due to the particular situation the requirements for the optimization may differ significantly. This makes optimization quite complicated, Fig. 3.

The layout-optimization has to deal with the selection and positioning of the appropriate production and logistic equipment. The case of planning new facilities is relatively seldom,

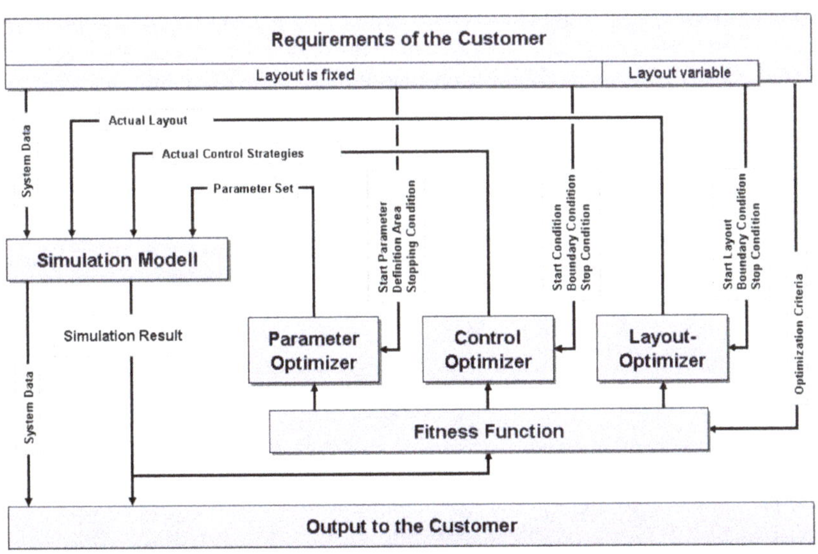

Figure 3: Optimization of parameter settings, control strategies and layout variants.

in most cases optimization space is limited by existing resources and given boundary conditions. Major layout or structural changes in existing systems effect significant effort and investment costs.

The control optimization focuses on the improvement of the operative production and logistic control, such as scheduling and routing of orders, and technical control strategies as well. Depending on the control architecture the optimization of control strategies may require changes in the software during the ramp up phase or even during operations.

The parameter optimization is related to production parameters, such as number of work piece carriers. In most cases the implementation of parameter changes is possible with relatively small effort even during the operational phase.

All these optimization levels are interconnected and changes on one level have always an impact on the other levels. The optimization on a single level will lead to a sub-optimum. In order to improve the digital enterprise regarding factory and systems layout, control strategies and parameter settings there are complex multi criteria decision making processes required. However, optimization in the academic sense assuming that a global optimum has to be reached is not realistic in industrial practice.

The solution space for an optimization is mostly limited. A sub-optimum, which provides good or at least satisfactory results and can be reached with relatively low expense regarding time and money, may be the better solution. Due to existing boundary conditions and the relatively high investment costs for layout changes, in most industrial cases, only parameter or control changes are considered.

3 DIGITAL TWIN CONCEPT

The goal of Digital Twins is to monitor and improve equipment and processes in a virtual environment. Digital Twins are virtual clones of real assets or processes. As the physical source is changing, the data from the physical asset or process are collected in real-time and replicated into the virtual equivalent in order to improve operative decision making [12]. A Digital Twin can help to improve product and processes continuously.

With the technology of IoT Digital Twins can be efficiently linked to real-world objects and offer information of the state of the assets responding to changes [13]. In production and logistic enterprises Digital Twins will enable to improve operations, asset management, operational efficiency, maintenance repair and operation and insights into how products and

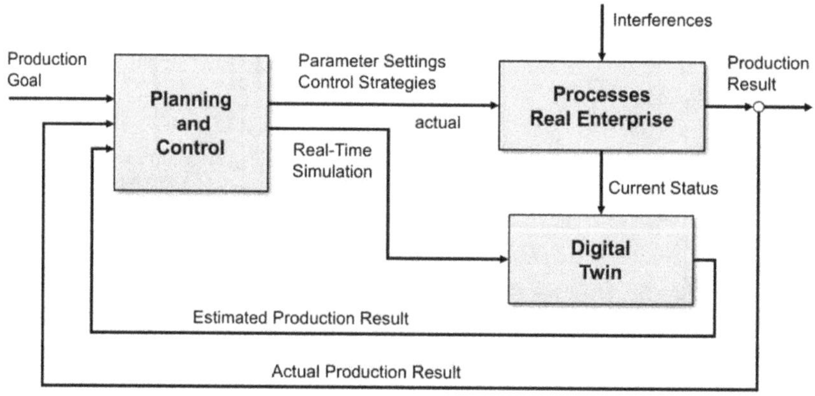

Figure 4: Digital Twin Approach

processes can be improved [14]. The monitoring and analysis of real-time data and advanced product, production or logistic simulation models allow to improve design, controls and strategies [15]. By use of simulation models the Digital Twins can be used to operate the enterprise or parts of it in advance and to test drive several alternatives in a virtual environment before a decision is applied to the real-world system [16]. With Artificial Intelligence based capabilities Digital Twins offer for specific questions advanced product, production or logistic simulation models, which allows to test drive several alternatives in a virtual environment before decisions are applied to the real-world system [17].

3.1 Closed Loop Approach

The Digital Twin approach concept in Fig. 5 contains an interaction of six steps which completes a closed loop connection (physical-to-digital-to-virtual-to-physical) between the physical world and the virtual model of the Digital Twin [19].

Create – With multiple sensors various inputs from the physical process and its environment are measured by use of integrated smart components. These measurements can be classified into the operational measurements sensing physical performance criteria of the asset and the measurement of environmental or external data affecting the operations of a physical asset. Signals from the sensors may be augmented with process-based information from systems such as the manufacturing execution systems, enterprise resource planning systems, CAD models and supply chains systems in order to provide the Digital Twin with a wide range of continually updating information to be used as input for the analysis.

Communicate – Network communication enables a seamless real-time connectivity between the physical process and the digital platform. Edge processing enriches sensors and process data with business and contextual data from the environment and passes the relevant data along to the platform. Proprietary protocols are translated into standard data formats. Network technology has been enabling the Digital Twin in practice. However, with

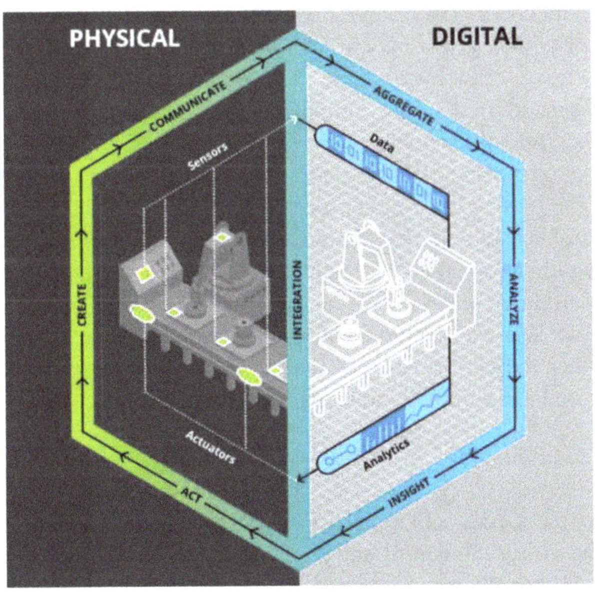

Figure 5: Manufacturing process Digital Twin model (*Source: Deloitte, 2017*)

an increasing number of Iot devices security issues increase and strong security approaches are required.

Aggregate – The real-time data have to be send into a data repository, processed and prepared for the analytics. The technologies of data aggregation and processing have evolved tremendously over the last years and allow nowadays to create massively scalable architectures with greater agility.

Analyse – The aggregated data are analysed by use of advanced analytics technologies in order to analyse that data on an ongoing basis to identify opportunities for possible improvements.

Insight – Based on the analysed data, models for decision making are created. Significant differences in the performance of the Digital Twin model and the physical world indicate potentials for improvement. Simple Digital Twin models visualizes these differences in order to enable recommendations. More advanced Digital Twin models use iterative simulation to analyse with the virtual model the possible impact of changes for the real world in order to generate insights for decision making.

Act – The knowledge and recommendations from the insights step can be fed back to the physical world in order to transform the real enterprise. The insights may feed directly into actuators for movement and control, or into software systems on a higher level for operational or supply chains behaviour. With this interaction Digital Twins complete a closed loop connection from the physical world to the virtual model and back to the physical world.

These six steps are easy to list. However, it requires significant effort to achieve. Companies that embrace Digital Twins have the opportunity to better understand and continuously improvement of products, services and processes, which gives them competitive advantage [19].

3.2 Interfacing Physical and Digital World

For the integration in digital enterprises sensor are absolutely essential for interfacing the real word. Sensor enable to monitor and analyse multiple machine variables to improve performance, reliability, and energy efficiency of the assets. Sensors can deliver vital insights into machine state and health, e.g. indication when bearings require lubrication or are wearing out. With modern sensor technology it is possible to affordably collect huge amounts of data from hundreds of sensors in a single equipment, analyse that data in real time and detect problems before they actually happen.

Sensor technology is moving from single sensor to integrated sensor systems as devices in IoT frameworks. The performance of sensor systems has increased significantly and at the same time the cost of sensors, network hardware, computing power, data storage, and communication bandwidth have fallen dramatically. Modern integrated electronic hard- and software and standard interfaces have made powerful sensor applications cheaper and faster to implement at scale. Wireless communication and handheld or wearable devices have made access easier. Further the performance of data-analysis, even already integrated in sensor systems, has increased and is moving towards Artificial Intelligence techniques.

3.3 Data Quality and Reliability

Data are the key for the success of Digital Twins and any Artificial Intelligence use case. It is crucial to determine the type, quantity, and quality of data. Real-Time Data are the key factor for Digital Twins. In recent years, real-time data generated across the manufacturing value chain have grown dramatically in volume and variety. These data are coming directly from smart products, connected production equipment, core manufacturing processes, enterprise IT systems, and external sources from customers or suppliers.

Digital Twins require real-time data from the physical systems in order to interface between the real and the virtual world. An important issue is the quality and reliability of these data gathered from various sensors. Therefore, in industrial application Digital Twins have to apply anomaly detection of sensor data. Using a combination of physics-based knowledge and collected sensor data Digital Twins may use five types of technologies for anomaly detection available [15].

Domain or physics-based methods – Physics based models enable anomaly detection through the comparison of calculated parameters with measured values. By analyzing the expected values throughout the physical system with the reported values, it is important to determine whether a true physical anomaly is occurring or if a sensor is out of calibration. If a sensor issue is identified the physics-based models can be used to provide a virtual sensor reading based on the remaining valid data. If a true anomaly is detected, then the Digital Twin applications shall provide recommendations to minimize potential impacts.

Advanced signal processing – To detect anomalies in the presence of sensor noise advanced signal processing techniques can be used. Algorithms using wavelets, kernel regression and multi sensor data fusion techniques may be applied.

Detection by use of Statistical process control – Univariate and multivariate control chart techniques can be used with thresholds set empirically or from domain expertise. These include algorithms to detect operational regimes to apply appropriate filters and control limits.

Machine Learning anomaly detectors – Machine learning algorithms can be used, ranging from multivariate multi-level survival models to baseline asset risk, to classification techniques like logistic regression, decision trees, random forest methods, neural networks and clustering methodologies. These models are usually derived using healthy and fault data based on a database of historical sensor and configuration data.

Deep Learning Anomaly Detection – Artificial Intelligence technologies like deep learning neural networks can be trained to detect anomalies.

3.4 Virtual Models and Automated Model Generation

For Digital Twins it is meaningful to create lightweight virtual models by selecting required geometry, characteristics, and attributes without carrying around unnecessary details. This reduces the size of the models and allows for faster processing. Light-weight models allow to simulate complex systems, including their physical behaviours, in real-time and with acceptable costs. The time and cost of communicating is substantially less and these models can be shared within the enterprise and throughout the supplier network. This enhances collaboration, understanding and quality of product and process information.

For Digital Twins an automated model generation can be performed based on real-time data. This model generation allows a rapid and flexible modelling approach also of large production areas. The great advantage of this approach is that the model is directly generated from the current data automatically and the state of the data and the simulation model matches always. If necessary specific details may be modelled by use of predefined specific building blocks.

3.5 Digital Thread

A digital thread can be woven through all areas of the digital enterprise from the customer to the supplier, integrating product lifecycle management, enterprise resource planning, manufacturing operations management to customer relationship management.

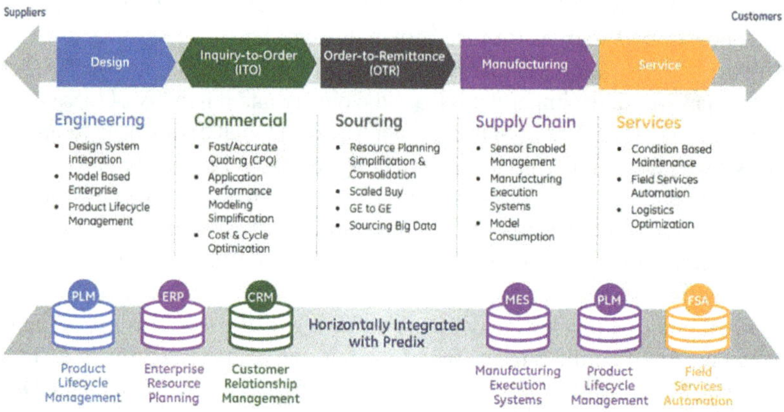

Figure 6: Digital Thread (*Source: General Electric, 2016* [15])

Nowadays increasing computation power and storage possibilities of big data engines, the improvement of analytics technologies and the integration of various data enables Digital Twins to model much richer, less isolated and much more sophisticated and realistic models than ever before.

A Digital Twin conceptual architecture has to be designed for flexibility and scalability in terms of the number of sensors and messages, applied analytics and processing in order to enable the architecture to evolve rapidly with growing demands [16].

4 ENTERPRISE DECISION MAKING

Decision making is becoming a very important key-factor in digital enterprises. Due to the complexity of modern enterprises the operation and optimization of complex production and logistic enterprises requires data and simulation driven multi criteria decision approaches.

4.1 Requirements

Digital enterprise systems are complex systems, in terms of layout variability, complex control strategies, business processes and system parameter. All of these aspects are not independent and even due to the system dynamics the optimal solution may differ depending on the actual situation and specific requirements.

Decision making can be classified in decision phases and decision levels. Due to circumstances, that in digital enterprises the decision making doesn't depend on a single parameter only, multi criteria decision making is required. Complex enterprises require a high flexibility and fast, clear and save decisions.

4.2 Rule-based Decision Making

In complex production and logistic enterprises with a high number of different products and processes, it must be decided frequently, how jobs shall be scheduled and routed through the facilities in order to fulfil the required goals at a maximum. This planning can be generated by a rule-based scheduling strategy. A particular strategy can be built up from several priority classes, filtering and sorting rules, that are defined as default rules in order to optimize the global system or at least to improve local requirements. Examples for such rules are e.g. "Earliest Due Date", "Critical Ratio" (Job with the least time to due date divided by total remaining processing time) or "Critical Path".

The rules can be used as a single rule or as a network of combined rules in order to consider different aspects. Today, most companies are using enterprise software with rule-based decision making and this rule-based enterprise software has brought tremendous productivity improvements. Because enterprises are very complex and due to a relatively long reaction time and the dynamic of the systems the effects of some rules are very difficult to oversee. Therefore, in the daily business, production planners implement often only a very few of the available rules for the daily operation. Digital Twins offer a powerful approach to test various strategies and modifications without risk for the current production.

However, rule-based decision software is limited and cannot learn and improve with experience as humans can. With modern computer technology more advanced approaches will be within reach.

4.3 Artificial Intelligence

Decision making can be improved significantly by use of Artificial Intelligence technologies [20]. Due to advances in computer processing power, nowadays machine learning can be integrated into the decision-making processes implemented into enterprise software systems [21]. By use of advanced decision-making approaches the software will be enabled to contribute increasing levels of performance and productivity.

Advanced Digital Twins employ Artificial Intelligence technologies that leverage data from equipment to generate insights and deeper understanding of operating environments. These includes unstructured data analytics, multi-modal data analytics, component analytics, pattern recognition, learning models, knowledge networks etc. [22]. One sub-field of AI is machine learning.

4.4 Machine Learning

Machine Learning (ML) is an application of Artificial Intelligence in which computers perform tasks by learning from data instead of being explicitly programmed. Machine learning uses algorithms to learn from data. The more data the algorithms can access, the more the algorithms can learn. Nowadays real-world machine learning examples are everywhere, such as personalized product recommendations from Web-sites, facial recognition, or fastest route suggestions in routing applications. Machine learning is still in the early stages of industrial development and implementation, but its potential is enormous. Machine learning has the potential to discover new signals in the data that could allow for continuous improvement of complex systems.

Machine learning is based on data, either from structured data and increasingly also from unstructured data without explicitly being programmed. Based on data and algorithm it is referred to as supervised learning, where the algorithm is trained using examples where the input data and the correct answers are known and in unsupervised learning, where the algorithm must discover patterns in the data on its own, and reinforced learning, where the algorithm is rewarded or penalized for the actions it takes based on trial and error [23].

Digital Twins can benefit from machine learning. Very much potential is in applications that have complex rules and large amounts of data. There are some basic requirements for machine learning [24]:

Clearly formulated issue – Machine learning needs a specific, well defined task where the relevant inputs and desired output can be clearly stated. Machine learning can be valuable in discovering correlations in large amounts of data that humans could never deduced for themselves. However, it is mandatory to focus on specific tasks, it is not possible that organizations learn from all the data in the enterprise just automatically.

Meaningful differences within the dataset – Data are always the main key-factor for machine learning. Machine learning algorithms will provide useful results only, if the data to learn from contain significant meaningful differences.

Sufficient quantity of examples to learn from – Machine learning requires a lot of data. There must be enough examples to learn meaningful approximations of the decisions to be performed.

Clear definition of success – Machine learning is always evaluated by measures of performance on specific tasks. Clear evaluation criteria for the algorithm are absolutely mandatory and it is very important, that the evaluation criteria really focus on the actual problem [25].

5 VALUE OF DIGITAL TWINS

The Digital Twin concept is very powerful and typical benefits of this concept are:

Visibility – The Digital Twin allows visibility in the operations of an asset, an interconnected system or even a complete enterprise.

Predictive – Using various modelling techniques, Digital Twin models can be used to predict future states of an assets.

What if Analysis – Through properly designed interfaces, it is possible to interact with the model and ask what-if questions to the virtual model and to simulate various conditions that are impractical to create in real life.

Understand and explain behaviours – Digital Twin model can be used as a communication mechanism to understand and explain the behaviours of an individual asset, machine or system.

Connect disparate systems such as backend business applications – Digital Twin models can be used to connect with the backend business applications to achieve business outcomes in the context of supply chain operations including manufacturing, procurement, warehousing, transportation and logistics, field service, etc. [12].

Digital Twins offer for specific questions an advanced product, production or logistic simulation model, which allows the monitoring and analysis of real-time data to improve design, controls and strategies. Digital twins offer a possibility to operate the enterprise or parts of it in advance and to test drive several alternatives in a virtual environment before a decision is applied to the real-world system. However, the digital twin concept requires a strong interface between virtual and real world by use of real-time data. The continuous improvement is increasingly important to operate complex production and logistic enterprises on a long run efficiently and successfully. The design and realization of Digital Twins requires a powerful combination of digital plus industrial strength, a combination of deep physics knowledge, engineering design knowledge, advanced sensing and inspection technologies expertise, Artificial Intelligence and analytics experience.

REFERENCES

[1] Menard, A., How can we recognize the real power of the Internet of Things? *McKinsey*, available at www.mckinsey.com/business-functions/digital-mckinsey/our-insights/how-can-we-recognize-the-real-power-of-the-internet-of-things, 2018 (accessed 4 June 2018)

[2] Gupta, V. & Ulrich, R., How the internet of things how the internet of things will reshape future production systems. *McKinsey* (Modern vision), available at www.mckinsey.com/business-functions/operations/our-insights/how-the-internet-of-things-will-reshape-future-production-systems, 2017 (accessed 4 June 2018)

[3] Alsén, D., Patel, M. & Shangkuan, J., The future of connectivity: Enabling the Internet of Things. *McKinsey*, available at www.mckinsey.com/global-themes/internet-of-things/our-insights/the-future-of-connectivity-enabling-the-internet-of-things, 2017 (accessed 4 June 2018)

[4] Deloitte, Industry 4.0. 2017 Global Impact Report, available at www2.deloitte.com/content/campaigns/global/global-report/global-report.html, 2017 (accessed 4 June 2018)

[5] Deloitte, Forces of change: Industry 4.0. Deloitte Development LLC, available at www2.deloitte.com/content/dam/insights/us/articles/4323_Forces-of-change/4323_Forces-of-change_Ind4-0.pdf, 2017 (accessed 4 June 2018)

[6] Parrott, A. & Warshaw, L., Industry 4.0 and the digital twin. Manufacturing meets its match. *Deloitte*, available at www2.deloitte.com/insights/us/en/focus/industry-4-0/digital-twin-technology-smart-factory.html, 2017 (accessed 4 June 2018)

[7] Bauer, H. & Burkacky, O., Security in the internet of things. *McKinsey*, available at www.mckinsey.com/industries/semiconductors/our-insights/security-in-the-internet-of-things, 2018 (accessed 4 June 2018)

[8] Deloitte, The Fourth Industrial Revolution is here – are you ready? *Deloitte Development LLC*, available at www.forbes.com/forbes-insights/wp-content/uploads/2018/01/Deloitte-FourthIndustrialRev_REPORT_FINAL-WEB.pdf, 2018 (accessed 4 June 2018)

[9] Siemens, Digital innovation. business transformation to realize innovation. *Siemens*, available at www.plm.automation.siemens.com/global/de/our-story/vision-digital-innovation/, 2018 (accessed 4 June 2018)

[10] Dassault Systèmes, The digital manufacturing enterprise: Leveraging ISA-95 as a foundation. *Dassault Systèmes*, available at www.apriso.com/library/Whitepaper_Digital_Manufacturing_Enterprise.php, 2017 (accessed 4 June 2018)

[11] Osan, A. & Somers, K., Optimizing production in the age of the machine. *McKinsey* (Modern vision), available at www.mckinsey.com/business-functions/operations/our-insights/optimizing-production-in-the-age-of-the-machine/, 2017 (accessed 4 June 2018)

[12] General Electric, The digital twin. Compressing time-to-value for digital industrial companies. *General Electric Company*, available at www.ge.com/digital/sites/default/files/The-Digital-Twin_Compressing-Time-to-Value-for-Digital-Industrial-Companies.pdf, 2016 (accessed 4 June 2018)

[13] Parris, C.J., Laflen, B., Grabb, M.L. & Kalitan, D.M., The future of industrial services: the digital twin. *General Electric Company*, available at www.infosys.com/insights/services-being-digital/Documents/future-industrial-digital.pdf, 2016 (accessed 4 June 2018)

[14] Oracle, Digital twins for IoT applications. A comprehensive approach to implementing IoT Digital Twins. Oracle (White Paper), available at www.oracle.com/us/solutions/internetofthings/digital-twins-for-iot-apps-wp-3491953.pdf, 2017 (accessed 4 June 2018)

[15] General Electric Company, GE digital twin. Analytic engine for the digital power plant. *General Electric Company*, available at www.ge.com/digital/sites/default/files/Digital-Twin-for-the-digital-power-plant-.pdf, 2016 (accessed 4 June 2018)

[16] Shetty, S., How to use digital twins in your IoT strategy. *Gartner*, available at www.gartner.com/smarterwithgartner/how-to-use-digital-twins-in-your-iot-strategy/, 2017 (accessed 12 March 2018)

[17] Chistty, P., Prepare for the impact of digital twins. *Gartner*, available at www.gartner.com/smarterwithgartner/prepare-for-the-impact-of-digital-twins/, 2017 (accessed 4 June 2018)

[18] Parrott, A. & Warshaw, L., Industry 4.0 and the digital twin. Manufacturing meets its match. *Deloitte*, available at www2.deloitte.com/insights/us/en/focus/industry-4-0/digital-twin-technology-smart-factory.html, 2017 (accessed 4 June 2018)

[19] Cline, G., Product development and the centrality of digital twin. *Aberdeen Group*, available at www-01.ibm.com/common/ssi/cgi-bin/ssialias?htmlfid=WWL12391USEN&), 2017 (accessed 4 June 2018)

[20] McKinsey, An executive's guide to AI. McKinsey, available at www.mckinsey.com/business-functions/mckinsey-analytics/our-insights/an-executives-guide-to-ai, 2017 (accessed 4 June 2018)

[21] Schick, U., What is artificial intelligence?, available at www.news.sap.com/what-is-artificial-intelligence/, 2018 (accessed 4 June 2018)

[22] Gramatke, M., Gegner, J., Bauer, S., Boenisch, M. & Le, V.D.B., Cognitive artificial intelligence the invisible invasion of the media business. *Deloitte*, available at www2.deloitte.com/de/de/pages/technology-media-and-telecommunications/articles/cognitive-artificial-intelligence-in-media.html, 2017 (accessed 4 June 2018)

[23] McDermott, B, Machines can't dream. *SAP*, www.news.sap.com/impact-of-artificial-intelligence-machines-cant-dream/, 2018 (accessed 4 June 2018)

[24] Wellers, D., Woods, J., Jendroska, D. & Koch, C., Why machine learning and why now?, available at www.sap.com/documents/2017/05/de7cfb6d-b97c-0010-82c7-eda71af511fa.html, 2017 (accessed 4 June 2018)

[25] Kolbjørnsrud, V., Amico, R. & Thomas, R.J., How artificial intelligence will redefine management. *Harward Business Review*, available at www.hbr.org/2016/11/how-artificial-intelligence-will-redefine-management, (accessed 4 June 2018)

FORMAL DESCRIPTIVE STUDY FOR THE EXTRACTION AND COMPARISON OF TOURIST SPENDING PATTERNS

A. RABASA, N. MOLLÁ-CAMPELLO & A. PÉREZ-TORREGROSA
Operation Research Center, University Miguel Hernández of Elche, Spain.

ABSTRACT

This paper presents the design of an in-depth descriptive analysis of data collected from public surveys at tourist information points. It uses a dataset that compiles different information related to trips to the Valencian Community (Spain). The aim of this study is to describe the patterns (association rules) that a certain type of expense has, and how this could be used to improve the services offered to tourists.

There are different kinds of expenses to analyze: transport, accommodation, leisure as well as total daily expenses and total daily expenses per person. Those cases where expenses are especially high or low are considered as particularly important because of their strategic interest for the public administration of tourism.

The study starts with data preprocessing, followed by pattern extraction for the sub-samples with very high and very low expenses, and in some cases, zero expenses are not considered as outliers but as a particular group of individuals. After this, the study aims to extract the most important attributes (feature selection) to create a classification model and compare its efficiency with the models that compute the complete set of attributes.

To conclude, this paper presents the possible future predictive models that could lead to an improvement in planning for public tourist services in the Valencian Community (Spain).
Keywords: Feature Selection, Pattern Discovery, Predictive Tourism Analysis.

1 INTRODUCTION AND OBJECTIVES

As the authors have illustrated in a previous study [1], the Valencian Community (Spain) is one of the main tourist destinations for foreign visitors. It is also an area where the tourism offer is growing and diversifying enormously. In that paper, the authors examined tourist spending segmentation from a descriptive approach, by using clustering models that grouped tourist spending according to their travel motivation (leisure, business…), transport and accommodation. It provided very useful information for tourist managers in order to help them to design accurate tourism offers.

In the same application framework, this paper examines a predictive approach, trying to create qualitative models that can classify tourist spending according to their corresponding travel patterns, which are extracted from public surveys (EGATUR) [2].

There are several classification algorithms (Table 1) that can be characterized with different splitting criteria and admit different attribute type or different pruning strategy.

CART [3] is one of the most applied algorithms in the classification field (discrete response or target variables) and regression (continuous target variables). One important aspect of CART is that it can do attribute selection for splitting and selecting those variables that provide higher information gain (Gini Entropy). However, different attribute selection techniques have been applied successfully in the tourism sector, such as Multiple Criteria Decision Making for hotel sites [4], Principal Component Analysis [5] or Support Vector Machine on Tourism Recommendation Systems [6].

In this paper, we raise the following question: from a descriptive model (always required) for pattern search, is it possible to extract a set of attributes that could improve the model's accuracy or obtain the same accuracy as those that use all the variables?

DOI: 10.2495/DNE-V13-N3-272-280

Table 1: Some decision tree models

	Splitting Criteria	Attribute type	Pruning Strategy
ID3	Information Gain	Only Categorical value	No implemented
CART	Gini Entropy	Categorical and Numeric value	Cost-Complexity
C4.5	Gain Ratio	Categorical and Numeric value	Error Based

These predictive models will be completely oriented towards establishing the extreme expense values (minimum and maximum), which are the most relevant aspects concerning public managers in the tourism sector.

2 DATA SET AND METHODOLOGY

2.1 Data Set

After data pre-processing and erasing the outliers, the global data set used for this study has a total number of 22742 records with 285 variables. From this global data set, several target variables are selected, including expenses on leisure, transport, accommodation or total expense, and a concrete data set is created for each expense using the minimum and maximum discretized ranges for each one. Thus, different files are created with different target variables (expenses on leisure, accommodation, transport, total), which are used to analyse the selected characteristics for each case.

Several subsets are created using different criteria to establish the maximum and minimum values for each type of expense (Table 2).

2.2 Methodology

The algorithms used for this study are CART and A priori, combined in order to improve the performance of the Classification And Regression Trees (CART) algorithm and to define the most important characteristics to establish extreme expenses (minimum and maximum expenses).

The generated subsets for expenses are trained and classified by the CART algorithm and obtain models with original accuracies, which will be compared with the accuracy of the new models trained by the CART algorithm with a previous attributes selection (using a priori algorithm). Thus, the methodology proposes to extract the 100 most reliable association rules from the minimum values and the 100 most reliable rules from the maximum values for each

Table 2: Summary of maximum and minimum levels of expenses

Expense	Criteria min	Criteria max	Values min	Values max	total minmax subset (% dataset)
Transport	0€–10€	> 100€	3877	2414	6291 (27.66%)
Accommodation	No expense	> 40€	11335	4816	16151 (71.02%)
Leisure	0€–33€	59€–80€	4460	4728	9188 (40.4%)
Total per day	0€–60€	> 200€	4117	4025	8142 (35.8%)

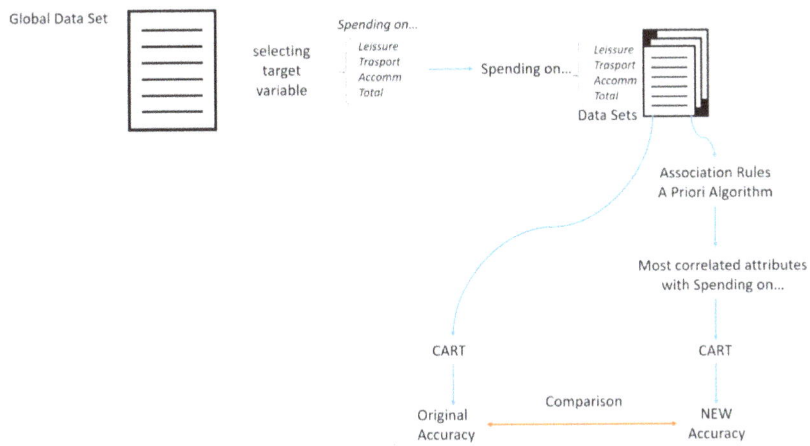

Figure 1: Summary of methodology

type of expense with the a priori algorithm. Those attributes that appear the greatest number of times in those rules are considered the most correlated variables with that type of expense and are selected to be trained by the CART algorithm in a new model, whose accuracy will be compared with the accuracy of the initial model.

3 COMPUTATIONAL EXPERIMENTS

The empirical results consist of performance estimates of the 2 classifiers across the data sets in terms of the four performance measures. Interested readers can find these raw results in Tables 2–10 and in the summary in Table 11.

3.1 Spending on leisure

Classification models that consider all attributes obtain an accuracy rate of 42.95%, a Type I error of 23.35%, a Type II error of 33.70% and root mean square error of 1.23. Table 3 shows the full confusion matrix.

The most frequent attributes in the association rules for leisure expenses (Fig. 2) are the person who paid the transport for the journey (GT2_4_2).

Table 3: Confusion Matrix for original leisure classification model.

	Very Low: 0–33	**Low: 34–43**	**Middle: 44–58**	**High: 59–80**	**Very High: >80**
Very Low: 0–33	840	306	200	82	33
Low: 34–43	76	176	68	4	6
Middle: 44–58	311	706	738	392	237
High: 59–80	55	101	190	458	265
Very High: >80	56	84	238	482	718

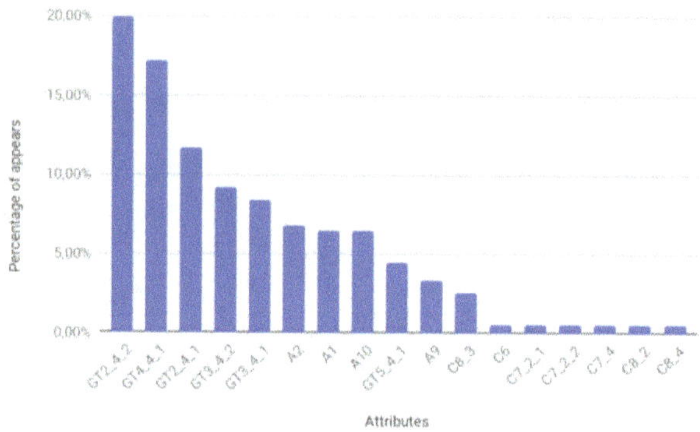

Figure 2: Ranking of variables for leisure expenses

The new Classification model results are: an accuracy rate of 74.20%, a Type I error of 2.27%, a Type II error of 23.53% and a root mean square error of 0.51 with only 17 variables. Table 4 shows the full confusion matrix.

3.2 Spending on transport

Classification models (that consider all attributes) obtain an accuracy rate of 61.65%, a Type I error of 20.71%, a Type II error of 17.63% and root mean square error of 0.93. Table 5 shows the full confusion matrix.

Table 4: Confusion Matrix for new leisure classification model.

	Very Low: 1–60	**High: >80**
Very Low: 1–60	727	59
High: >80	611	1200

Table 5: Confusion Matrix for original transport classification model.

	Very Low: 0–10	**Low: 11–20**	**Lower-Middle: 21–40**	**Higher-Middle: 41–60**	**High: 61–100**	**Very High: > 100**
Very Low:0–10	933	169	52	14	4	2
Low:11–20	210	633	263	29	22	22
Lower-Middle 21–40	3	160	1191	287	129	128
Higher-Middle: 41–60	0	0	177	456	127	26
High:61–100	0	1	76	352	586	139
Very High:> 100	17	5	4	22	176	407

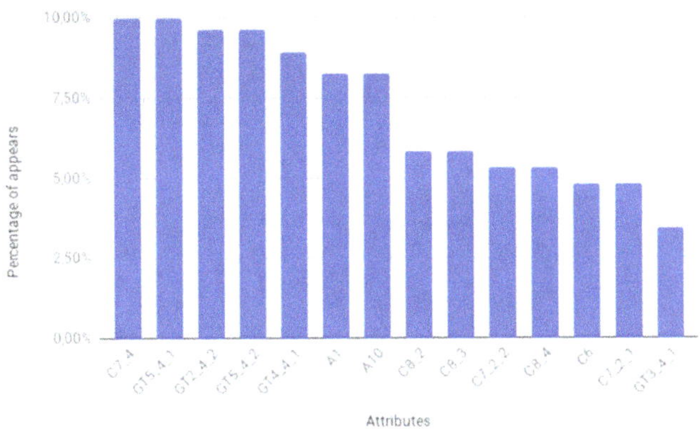

Figure 3: Ranking of variables for transport expenses

The most frequent attributes in the association rules that lead to minimum and maximum spending on transport are: the person who paid the tourist package (C7_4) and the person who paid a rented vehicle (GT5_4_1).

The new Classification model issues are: an accuracy rate of 91.89%, a Type I error of 0.26%, a Type II error of 7.84% and a root mean square error of 0.29 with only 14 variables. Table 6 shows the full confusion matrix.

3.3 Spending on accommodation

The classification model with all attributes obtains an accuracy rate of 83.65%, a Type I error of 8.66%, a Type II error of 7.68% and a root mean square error of 0.61. Table 7 shows the full confusion matrix.

The most frequent attribute in the association rules for spending on accommodation is clearly the reason for the trip (C1). Also, attributes like agency use (C10_1) or advanced booking (C10_2_1).

Table 6: Confusion Matrix for new transport classification model.

	Very Low: 0–10	**High: >100**
Very Low: 0–10	1015	5
High: > 100	148	719

Table 7: Confusion Matrix for original accommodation classification model.

	Equal to 0	**Low: 1–20**	**Middle: :21–40**	**High: > 40**
Equal to 0	3392	126	115	120
Low: 1–20	2	404	151	13
Middle: 21–40	0	74	665	66
High: > 40	6	11	431	1245

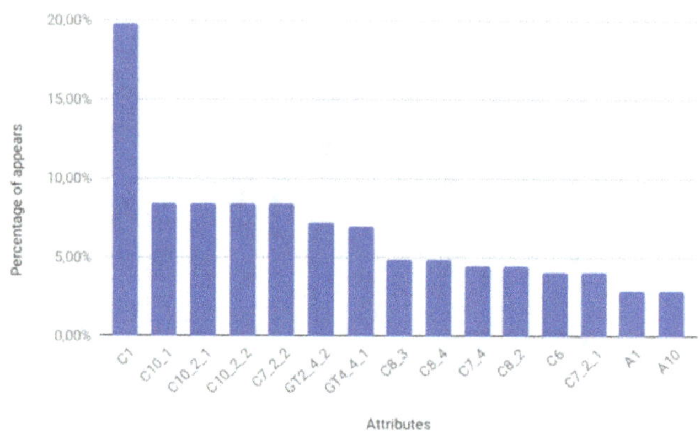

Figure 4: Ranking of attributes for accommodation

The new Classification model results are: an accuracy rate of 97.36%, a Type I error of 2.48%, a Type I error of 0.17% and a root mean square error of 0.16 with only 15 variables. Table 8 shows the full confusion matrix.

3.4 Total spending

Classification models that consider all attributes obtain an accuracy rate of 55.20%, a Type I error of 21.14%, a Type II error of 23.66% and root mean square error of 0.84 (Table 9).

The most frequent attributes are the following (Fig. 5):

Some examples for the most frequent attributes in the association rules that lead to minimum and maximum total expenses are: who paid the transport (GT2_4_2) or how many nights their stay is (A10).

Table 8: Confusion Matrix for new accommodation classification.

	Equal to 0	**High: > 40**
Equal to 0	3392	120
High: > 40	8	1324

Table 9: Confusion Matrix for spending

	Very Low: 1–60	**Low: 61–100**	**Middle: 101–130**	**High: 131–200**	**Very High: > 200**
Very Low: 1–60	666	180	8	3	1
Low: 61–100	554	1292	353	192	62
Middle: 101–130	14	149	175	122	25
High: 131–200	1	171	542	1009	496
Very High: > 200	0	1	11	171	623

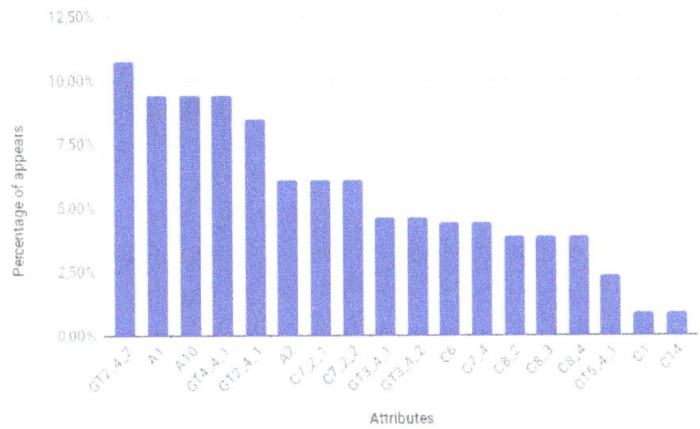

Figure 5: Rank by apriori attribute selector for Total spending

The new Classification model issues are: an accuracy rate of 85.38%, a Type I error of 5.81%, a Type II error of 8.80% and a root mean square error of 0.38 with only 18 variables. Table 10 shows the full confusion matrix.

Below, Table 11 summarizes the above results so as to compare the classification task improvement when completed.

Table 10: Confusion Matrix for new total spending classification model.

	Very Low:1–60	**High: > 200**
Very Low: 1–60	1020	142
High: > 200	215	1065

Table 11: Summary of different models

	ORIGINAL CLASSIF. MODEL (considering full set attributes)					NEW CLASSIF. MODEL (considering selected attributes)				
Spending on…	**var**	**Accu-racy**	**SME**	**Type I error (FP)**	**Type II error (FN)**	**var**	**Accu-racy**	**SME**	**Type I error (FP)**	**Type II error (FN)**
Leisure	52	42.95%	1.23	23.35%	33.70%	17	74.20%	0.51	2.27%	23.53%
Transport	52	61.65%	0.93	20.71%	17.63%	14	91.89%	0.29	0.26%	7.84%
Accommo-dation	52	83.65%	0.61	8.66%	7.68%	15	97.36%	0.16	2.48%	0.17%
Total	52	55.20%	0.84	21.14%	23.66%	18	85.38%	0.38	5.81%	8.80%

By using the A Priori Min-Max algorithm, we reduce the number of variables trained by the CART algorithm from 52 to 17 (leisure), 14 (transport), 15 (accommodation) and 18 (total), improving the accuracy associated to each model in all the cases.

4 CONCLUSIONS

In most cases, the results obtained improve considerably when the A Priori Min-Max algorithm (with minimum and maximum values) is used instead of the original attributes CART selection (with Gini index) trained against the whole dataset. Of course the binarization procedure provides such a high confidence ratio.

The main advantage obtained with the A Priori Min-Max method is that it guarantees an ordered methodology that allows the inclusion of a variable according to its relevance, thereby providing more control over the tree creation. Furthermore, the most frequent patterns are extracted in order to observe the evolution of the data.

The final variable sets chosen by the CART and the A Priori Min-Max algorithm have been evaluated by expert public managers in tourism. These agents confirmed that the variables chosen by the A Priori Min-Max algorithm (the most frequent attributes) have a greater strategic interest in most of typical study cases for any type of expense analysed [7], regardless of the particular sample used for the computational experiment.

5 FUTURE RESEARCH LINES

Improvement in accuracy in each case is due to binarization in extreme expenses and not so much to attribute selection. Thus, future studies will proceed to test other methods for attribute selection from datasets with prior binarization of extreme expenses [8].

On the other hand, the CART algorithm itself can be modified in terms of different information gain criteria, which necessarily leads to different attribute selections.

REFERENCES

[1] Rabasa, A., Pérez-Martín, A. & Giner, D., Optimal clustering techniques for the segmentation of tourist spending. Analysis of tourist surveys in the Valencian community (Spain): a case study. *International Journal of Design & Nature and Ecodynamics*, **12**(4), pp. 482–491, 2018. https://doi.org/10.2495/dne-v12-n4-482-491

[2] 'EGATUR', available at http://www.ine.es/dyngs/INEbase/es/operacion.htm?c=Estadistica_C&cid=1254736177002&menu=ultiDatos&idp=1254735576863 (accessed March 2018)

[3] Breiman, L. (ed), *Classification and regression trees*, Repr. Boca Raton: Chapman & Hall [u.a.], 1998.

[4] Aksoy, S. & Ozbuk, M.Y., Multiple criteria decision making in hotel location: does it relate to postpurchase consumer evaluations?. *Tourism Management Perspectives*, **22**, pp. 73–81, 2017. https://doi.org/10.1016/j.tmp.2017.02.001

[5] Nilashi, M., Bagherifard, K., Rahmani, M. & Rafe, V., A recommender system for tourism industry using cluster ensemble and prediction machine learning techniques. *Computers & Industrial Engineering*, **109**, pp. 357–368, 2017. https://doi.org/10.1016/j.cie.2017.05.016

[6] Pronoza, E., Yagunova, E. & Volskaya, S., Aspect-Based Restaurant Information Extraction for the Recommendation System. *Conference: 6th Language and Technology Conference. Lecture Notes in Computer Science*, 9561, pp. 371–385, 2016.

[7] Sinclair, M.T. & Stabler, M., *The Economics of Tourism*. Library of Congress Cataloguing in Publication Data, 2002.

[8] Pol, A.P., Pascual, M.B. & Vazquez, P.C., Robust estimators and bootstrap confidence intervals applied to tourism spending. *Tourism Management*, **27**(1), pp. 42–50, 2006. https://doi.org/10.1016/j.tourman.2004.06.016

A COMPLEXITY FRAMEWORK FOR CONSENSUS AND CONFLICT

PETER MITIC
Santander UK, 2 Triton Square, Regent's Place, London NW1 3AN
Dept. of Computer Science, UCL, Gower Street, London WC1E 6BT
Laboratoire d'Excellence sur la Régulation Financière, Paris

ABSTRACT

Consensus and conflict are modelled in the context of interacting pairs of agents who may have very diverging sentiments regarding some particular issue. Simulations using the model display characteristics of complexity. Agents are modelled using *Beta* probability density functions whose parameters determine the agent's opinion and resistance to change after an interaction, and a third independent parameter that determines the agent's influence. Interactions among groups of agents with both aligned and opposing sentiments are simulated. The results indicate that in most cases a form of consensus is reached eventually, but for opposed agents, it is not possible to tell which agents that consensus will favour. Proofs of convergence are given in the cases where the initial state is one of consensus, and when it is one of conflict.

Keywords: Beta distribution, conflict, consensus, convergence, influence, resistance, sentiment, simulation

1 INTRODUCTION

In this analysis we build a framework for complexity that is consistent with the characteristics of complexity stated by Rzevski in [1]: *autonomy, emergence, non-equilibrium, non-linearity, connectivity, self-organisation* and *co-evolution*. The framework, once initialised, is designed to evolve with time according to rules that are stochastically determined, and follow general principles. Those principles are consistent with Rzevski's discussion in [2], namely that a model of a complex system must be adaptable to changes, and that the adaptation must be autonomous (i.e. without specific instructions). A major consequence is that the agents in the system must be able to self-organise (see the details in Section 3).

The framework comprises a set of n agents $\boldsymbol{X} = \{X_1, X_2, \ldots, X_n\}$. In the interaction model of Section 3, time is discretised such that times $t_1, t_2, \ldots$ are the times at which an agent-agent interaction occurs. The resulting time intervals, $t_2 - t_1, t_3 - t_2, \ldots$ are therefore not necessarily of constant length. The framework is used to examine the general principles of conflict and consensus in the context of agreement or disagreement on some issue. We say that an agent has a particular *sentiment* with respect to an issue, and measure that sentiment by a real number in (0,1) which represents the degree to which the agent either agrees or disagrees with some proposition.

1.1 Conflict and Consensus

Any pair of agents can "agree" or "disagree" with each other with respect to an issue. Agreements constitute *consensus*, and large disagreements constitute *conflict*. The precise definition of an agent (Section 3.1) will clarify the distinction between *conflict* and *consensus* adequately. The development over time of opinions of a group of agents will be simulated (Section 4). The complexity-derived simulations determine whether or not consensus can be reached following initial conflict, and whether or not conflict can arise from initial consensus.

DOI: 10.2495/DNE-V13-N3-281-293

The interaction between agents is modelled by the balance of influence of one agent and resistance of another. That balance results in emergent behaviour: non-deterministic results which sometimes cannot be explained precisely.

2 PREVIOUS WORK ON CROWDS, CONFLICT AND CONSENSUS

This section contains a brief account of studies of agent interaction that relate to consensus or conflict. Early work in this context was done by Epstein [3], and it forms the basis of many subsequent studies. Epstein states principles, broadly the same as the ones stated in the previous section, in the form of interaction rules. However, no specific model is proposed, and *emergence* is not mentioned. In a later paper [4], Epstein introduces a rule-based model in which the state of a *non-authority* set of agents can be changed by an *authority* set of agents. The model incorporates hard thresholds: breaching a threshold results in a state change.

Epstein's principles persist in the review of Lemos *et al.* [5], with the addition of *emergence*. The context remains social conflict. They discuss, first, threshold-based models, such as from Granovetter [6]. Each agent has a threshold, and joins a protest if the number of other agents in the protest exceeds the threshold. Among other more specialised models cited are papers by Doran [7] (weak but agile agents against strong but sluggish agents), and Kim & Hanneman [8] (rules for opposing agents are distribution-based). In the context of threshold models, we also mention Gurr's *Relative Deprivation* model [9], in which an agent's increasing disparity between expectation and reality leads to a change of state.

Game theory is a second, and common alternative, approach to agent interaction. This choice is natural, since such models use constraints on actions due to other agents. Jones [10] gives a comprehensive overview of conflict and consensus models with non-cooperative and cooperative game theory respectively. Among the concepts considered are threats, bargaining and strategy. Axtell *et al.* [11] models agent interaction as a game where two players bargain for ownership of a resource. If their joint demand exceeds some predetermined limit, they both get nothing. Otherwise, they get their demands. All possible outcomes are shown, as is the location of each agent with respect to those outcomes. Weisbuch [12] extends the Axtell analysis to include features (e.g. a moving average of previous gains) for each agent. Weisbuch's result shows that there is a stabilisation as time increases, and that divergence does not emerge spontaneously from a random starting point, agreeing with results in Section 4 of this paper.

Game theory models tied to a network represent a third mode of agent interaction. Many have the *Voter* model [13] as a basis. Within a network, a *target* agent having a discrete set of opinions is selected at random, and is then assigned the opinion of one of its nearest neighbours, also chosen at random. In terms of the network topology context, the 'nearest' neighbours are the agents that are directly connected to the *target* agent. Physical distance between the *target* agent and any other agents is not relevant: network connections are virtual.

This opinion-formation mechanism is best suited as a model of agents' lack of self-confidence, resulting in imitation. Under assumptions that state transitions are binary (only two opinions, with a potential transition of one to another), and that such transitions follow a Markov process, agents' opinions converge to one of the possible opinions. The *Axelrod* model [14] is an extension of the *Voter* model in that each agent is assigned a set of features. In an agent-agent interaction, corresponding features are compared, and if they differ, the features of one agent are set to the features of the other. After multiple interactions, the *Axelrod* model predicts either global convergence, or the emergence of several groups. The *Axelrod* model is appropriate when applied to social groups that have multiple distinct characteristics, some of which are opposed. Simulations and summaries for both the *Voter* and *Axelrod* models are shown in Miguel *et al.* [15].

3 AGENTS AND AGENT INTERACTION

In this section we describe a new way to represent agents and to investigate interactions between them. This analysis differs from the prior analyses (Section 2) in that the context in which the agents interact is completely general. No assumptions are made apart from the general principles underlying complexity. The *Mathematica* implementation permits, a suitable combination of mathematical formulation, manipulation ability, and display. Example applications are given in this section, and in Section 4.

3.1 Agent Structure

An agent A is represented by a triple:

$$A = \{(a,b), I, s\}. \tag{1}$$

In (1), a and b are the parameters of a Beta density $\beta(a, b, x)$; $x, a, b \in (0,1)$, $I \in (0,1)$ represents the agent's influence and s is an alphanumeric name for the agent. With those values of a and b, a Beta density has a single peak with a skew that depends on the values of a and b. The degree to which an agent agrees or disagrees with an issue (i.e. the agent's opinion) is modelled by the location of the peak of the density in (0, 1). Specifically, an agent's location $L(A)$ is given by the mean of the density,

$$L(A) = \frac{a}{a+b}. \tag{2}$$

An agent with a biased opinion is represented by either $a < b$ (the peak is near $x = 0$) or $b < a$ (the peak is near $x = 1$). The parameter values in very biased cases are approximately in the ratios 1:10 and 10:1 respectively. An agent with a neutral opinion has $a = b$, and that opinion is most open to influence if a and b are small (typically less than 10). Both very biased and near neutral opinions are typically found in political contexts. Most recently very biased opinions have been noted with respect to referenda. Examples are Brexit, Scottish independence and Catalonian independence. In the illustrations in the following section, an agent with very low influence ($I \sim 0$) is shown with very low opacity, and an agent with very high influence ($I \sim 1$) is shown with very high opacity. Intermediate opacities indicate medium influence ($I \sim 0.5$). It is easy to identify people who are or have been influential. The result of their influence may only be apparent with hindsight. One example is Rosa Parks, who triggered the campaign for racial equality in the USA in 1955. Another is Nelson Mandela, who had a key role in formulating proposals to end apartheid in South Africa in 1989.

3.2 The Mechanism for Agent Interaction

The way in which agents interact is a more sophisticated version of the mechanism used by Burt [16]. The mechanism uses an agent's *resistance* to influence, ρ, which is defined in terms of the variance, v, of the agent's underlying Beta (a, b) distribution, ultimately mapped to (0,1) using a sigmoid transformation on the inverse square root of v (ρ' in (3), below).

$$v = \frac{ab}{(a+b)^2(a+b+1)};\ \rho' = \frac{1}{\sqrt{v}};\rho = \frac{2}{1+e^{-\frac{\rho'}{25}}} - 1 \tag{3}$$

When two agents $X((a_X,b_X),I_X,\text{"X"})$ and $Y((a_Y,b_Y),I_Y,\text{"Y"})$ interact, the convention used is that Y affects X, Y remains unchanged afterwards, and X transforms to a resultant, $Z((a_Z, b_Z), I_Z, \text{"Z"})$, denoted by $Z = <X, Y>$. The interaction proceeds by defining Beta parameters (a_Z, b_Z) for Z according to *Algorithm 1*, below.

Algorithm 1: Interaction $Z = <X, Y>$
Step 1: Calculate a linear combination of influences.

$$r_1 = \frac{2I_Y}{I_Y + \rho_Y} - 1; r = \frac{1}{1+e^{-2r_1}};$$

$$a'_Z = ra_X + (1-r)a_Y \; ; \; b'_Z = rb_X + (1-r)b_Y$$

Step 2: Modify the Beta parameters in Step 1 based on resistances, with a bias towards X.

$$a_Z = \sqrt{\rho_X}\,a_X + \left(1-\sqrt{\rho_X}\right)a'_Z \; ; \; b_Z = \sqrt{\rho_X}\,b_X + \left(1-\sqrt{\rho_X}\right)b'_Z$$

Step 3: Modify the Beta parameters in *Step 2* by perturbations (see the explanation below). Using single draws ε_{a_Z} and ε_{b_Z} from $N(0, a_Z/10)$ and $N(0, b_Z/10)$ distributions respectively,

$$a_Z \rightarrow a_Z + \varepsilon_{a_Z}; \; b_Z \rightarrow b_Z + \varepsilon_{b_Z}.$$

Step 4: Use a similar stochastic perturbation to derive the influence of Z, where ε_{I_X} is a single draw from a $N(0, I_X/100)$ distribution.

$$I_Z = I_X + \varepsilon_{I_X}$$

Step 5: Apply limiters such that the results of steps 3 and 4 are in (0,1)

The perturbation stages (3 and 4) add a non-linear element to the Beta parameter transformations. They make the interaction process non-deterministic, and model the situation where agents do not make entirely rational or expected decisions. An example of such a decision is the election of Donald Trump as US President. Overall, the effect of the *Algorithm 1* rules say: "Balance influence and resistance with a bias towards the opinion of the influenced agent." Figures 1a and b show two contrasting interactions using the *Algorithm 1* rules. They both

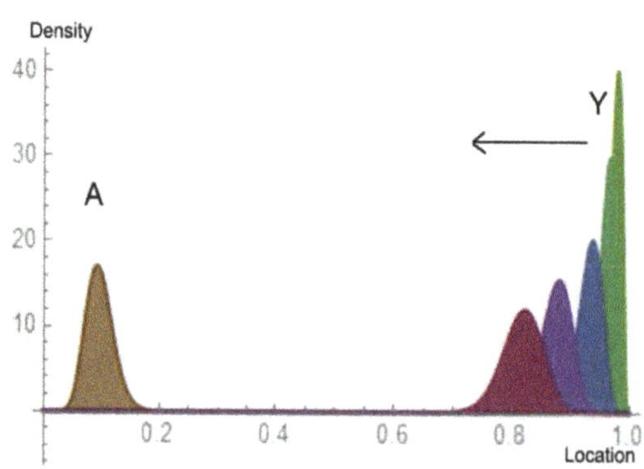

Figure 1(a): Influential A has a has a limited effect on resistant Y

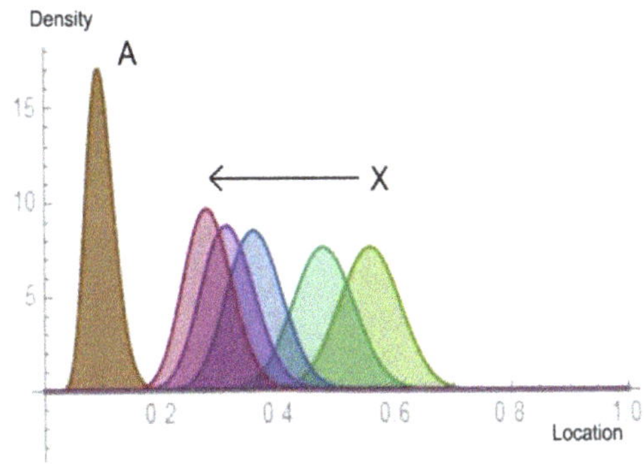

Figure 1(b): Influential *A* significant effect on non-resistant *Y*.

illustrate the repeated effect of an agent *A* a second agent. In Fig. 1a, the resultants of successive interactions are given by $Y_{r+1} = Y_r, A, r = 1..4; Y_1 = Y$. In this case, *A* is a very influential (high opacity) agent with high resistance, who tries to influence a *Y* who is equally influential and resistant, but whose opinion directly opposes *A*'s opinion. *A* has some success in shifting *Y*'s opinion, as shown by the gradual progression of *Y* from location near 1 to location near 0.8 (the left-pointing arrow). This is a model of, for example, unsuccessful advertising. In Fig. 1b, *Y* is replaced by an agent *X* who has medium influence (shown as lower opacity), has lower resistance to persuasion, and is not opinionated. *A* is able to shift *X*'s opinion significantly to a location near 0.3. Successive resultants in this case are given by $X_{r+1} = X_r, A, r = 1..4; X_1 = X$. This situation models successful advertising, and also successful transmission of influence by social media.

3.3 *Group Location* and Convergence

A *Group* is a set of agents. The *Group Location* is a weighted average of the locations of the members of a *Group*. The weights are normalised ratios of influence to resistance for each member of the *Group* (Equation 4). So for a Group $\boldsymbol{X}$ with *n* agents $\{X_1, X_2, \ldots, X_n\}$,

$$w_i = \frac{I(X_i)}{\rho(X_i)}, i = 1..n; L(\boldsymbol{X}) = \sum_{i=1}^{n} w_i L(X_i). \tag{4}$$

Following a sequence of agent interactions among members of a *Group*, the *Group Location* eventually converges to a consensus. The term 'convergence' is used in the sense that there exists a confidence interval in which the *Group Location* will lie following all interactions. Appendix A is a proof that, under certain conditions, the *Group Location* for the case $n = 2$ is bounded by a *subset* of (0,1). Appendix B extends this result to the case $n > 2$. In the case of agents whose opinions broadly agree, the eventual opinion will also broadly agree with the initial opinions. If two subsets are markedly opposed, there will be consensus eventually, but it is not possible to say which subset will be favoured, despite a bias towards influenced agents who are resistant to change (*Step 2* in *Algorithm 1*).

4 SIMULATION RESULTS

The principles introduced in the previous section are illustrated in this section. The appendices give proofs that predict what the results of simulations should be, and the simulations are consistent with those predictions.

Consider r interactions for the *Group* $\boldsymbol{X}$ of Section 3.3. At each interaction pairs of agents are chosen at random, with replacement. If the pair comprises two identical agents, the interpretation is that the agent changes its opinion independently of other agents. The emergent behaviour is then examined by generating pair-wise interactions repeatedly. The existence of an implied fully connected network with agents as nodes is assumed. Notable cases of influence and resistance follow. In each case, the plots presented show the *Group Location* directly after each interaction. In each case $n = 10$ and $r \geq 200$. Note that the interval between successive interactions is not necessarily constant. In each simulation, when agent $X_p \in \boldsymbol{X}$ tries to influence agent $X_q \in \boldsymbol{X}$, let the resultant after the rth interaction of N for X_q be $X_{[N,q,r]}$. Then $X_{[N,q,r+1]}$ is defined by Equation 5a. Equation 5b gives the *Group Location* $L(\boldsymbol{X}_{r+1})$ after r interactions (the weights w_i are defined in Equation 4).

$$X_{[N,q,r+1]} = \left\langle X_{[N,q,r]}, X_p \right\rangle, r = 1..N; X_{[N,q,1]} = X_q \tag{5a}$$

$$L(\boldsymbol{X}_{r+1}) = \sum_{i=1}^{N} w_i L(X_{i,r+1}) \tag{5b}$$

The *Group Locations* are illustrated in the figures that follow. If $p = q$, the interpretation is that X_q has influenced itself (i.e. has had a change of mind).

4.1 Consensus: Distinct Groups, each with Similar Opinions

Figure 2(a) shows three distinct *Groups* $G1$, $G2$ and $G3$ in their initial states, and Fig. 2b shows them after 200 repeated interactions. The 'before' and 'after' states look very similar. $G1$ and $G3$ are highly opinionated and remain clustered around their respective start locations, 0.9 and 0.1. $G2$ has no firm opinion and remains clustered near its starting location, 0.5. Figure 2(c) shows the corresponding *Group Location* for each *Group*. The $G1$ and $G3$ *Group Locations* are bounded within narrow limits, whereas $G2$ exhibits greater volatility (see Appendix A). These results reflect greater susceptibility of $G2$ to influence. These cases are typical of political groupings: some people are highly opinionated with respect to certain political parties, and they mostly stay that way. Others are willing to be swayed.

The results obtained in this simulation resemble those obtained by Namatame and Chen [17] in the context of reinforcement learning. They derive similar-shaped probability distributions in simulations, but have no explicit densities.

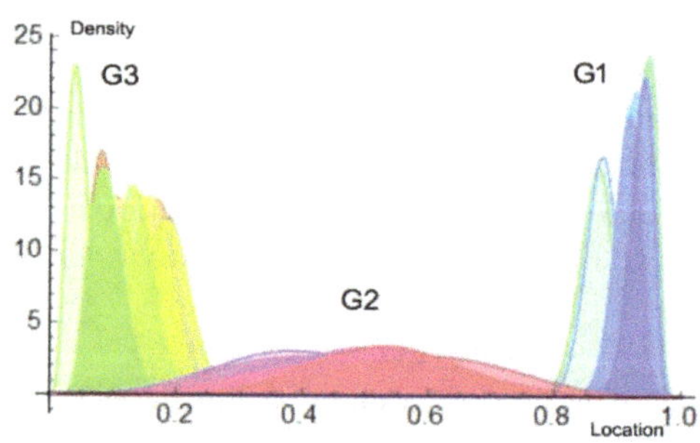

Figure 2(a): Three *Groups* initial state

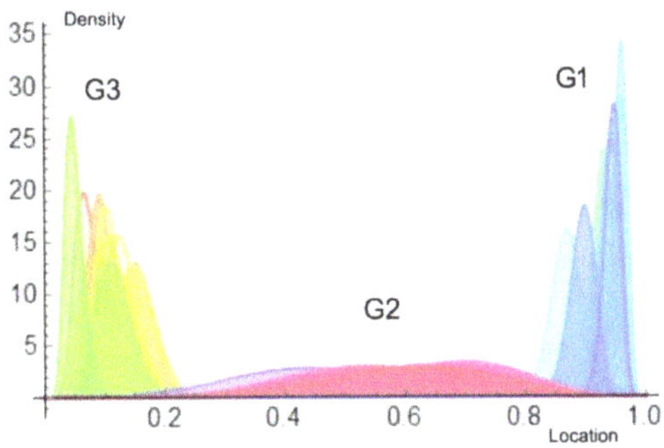

Figure 2(b): Three *Groups* final state

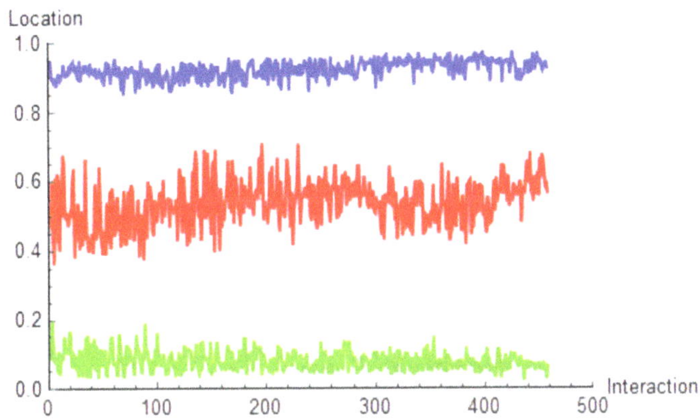

Figure 2(c): Three Groups: Time progression of Group Location (G3 in green, G2 in red, G1 in blue)

4.2 Conflict: Groups with Opposing Opinions

Figure 3 shows the progressive *Group Locations* of three distinct simulations for the same two agents, who have directly opposing strong opinions, as shown by their initial locations: 0.95 and 0.05. The simulations have common characteristics: initial volatility followed by local convergence (as predicted in Appendix B). However, the limit points are different for every simulation. If a simulation is run for 1000 interactions, the pattern of shifting the *Group Location* of the pair between the limits 0 and 1 continues. There is an eventual 'winner' of the argument, but it is not possible to tell in advance which agent will be the winner. This result is really the only one of the results presented here that is surprising. It is not clear, in advance of running a simulation, that any such consensus might be reached. A plausible expected result is continued conflict. The Anglo-Irish Good Friday agreement of April 1998 provides an example of where two conflicting parties eventually agreed on the issue of peace in Northern Ireland. Prior to that date it was not clear if an agreement would favour one side or the other, or to what extent.

4.3 The Effect of one Influential Agent on a Group

The illustration in this section (Fig. 4) shows the effect of a single influential agent, who joins a *Group* after the original *Group* has been interacting for some time. The entry point is at

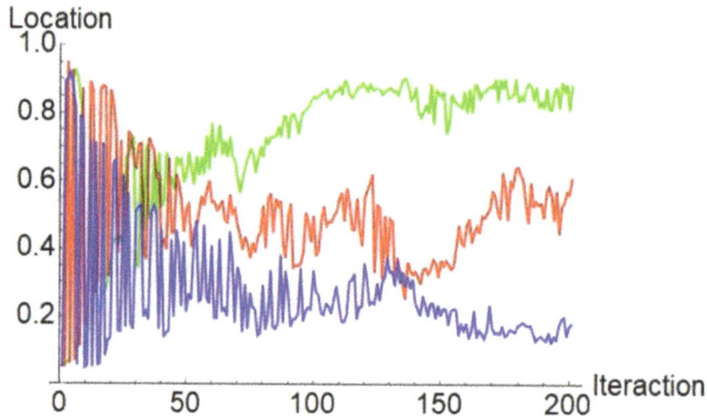

Figure 3: Three simulations: two opposed agents

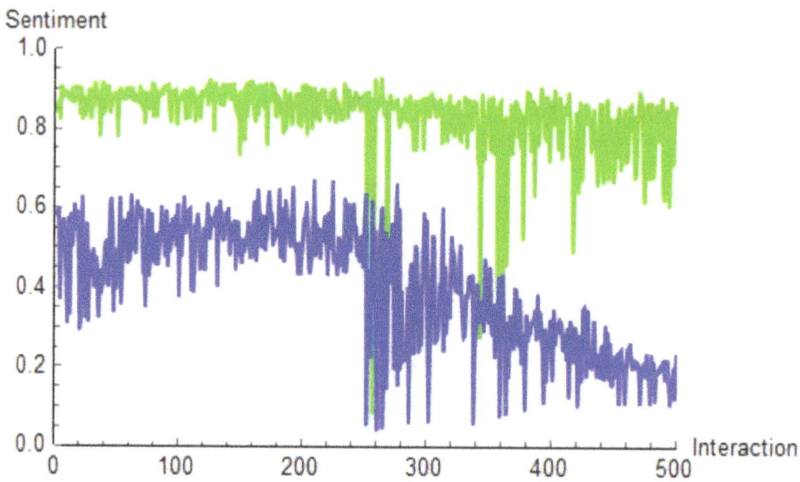

Figure 4: An opposing agent, entering at *Interaction* 250, having limited influence on a very biased *Group* (green trace), but more influence on a neutral *Group* (blue trace)

Interaction = 250: the half-way stage. Two traces are shown. The first represents a *Group* that holds a strong biased opinion, and the second represents a *Group* that holds an overall neutral opinion. The agent that enters has a strong opinion that opposes the first *Group*, and induces a period of volatility in the *Group Location* of both *Groups*.

If the *Group* has a hardened initial opinion, the opposing agent cannot really influence its members significantly. The dips in the *Group Location* show that the opposing agent has some effect, but there is no trend towards the opposing opinion for the original *Group*. As time progresses, the opposing agent is only able to generate volatility. Alternatively, the influencer can influence permanently the *Group* that was neutral initially. For such a *Group*, the *Group Location* trace moves towards the influencer's location, and stays there. There is inherently more volatility for a *Group* that is neutral initially. In both cases volatility increases markedly when the influencing agent enters. This simulation illustrates two points. First, the effect of a single influential agent on a *Group* can be significant. An analysis of the Brexit referendum (https://

www.referendumanalysis.eu/eu-referendum-analysis-2016/section-7-social-media/impact-of-social-media-on-the-outcome-of-the-eu-referendum/) indicates that the *Vote Leave* campaign had a significant effect on the outcome. Second, conflict can arise where there was initial consensus. Brexit provides an example here too: a rise in hate crime was widely reported following the referendum (https://www.independent.co.uk/news/uk/home-news/racist-hate-crimes-surge-to-record-high-after-brexit-vote-new-figures-reveal-a7829551.html).

4.4 Random Interactions with varying *Group* size

Two sets of randomly-defined agents (with 10 and 100 agents respectively) interact separately. In both cases their influences span the range (0,1) and their Beta distribution parameters are randomly chosen integers from the set 2..202. This allows for a comprehensive range of resistances and biases. We observe a much greater variation in the *Group Location* for the "100 agent" case (Fig. 5). Similar simulations show that as the *Group* size increases, so does the variation in the *Group Location*. The conclusion is that the more agents there are in the *Group*, the less likely it is that consensus will be reached. Large corporate organisations are examples where consensus is slow to emerge, whereas small startups are usually much faster. (https://www.fastcompany.com/3049164/how-to-make-decisions-more-efficiently).

5 DISCUSSION AND SUMMARY

The simulation results in Section 4 indicate that agents who 'agree' initially, continue to 'agree' (Section 4.1), a conclusion which makes intuitive sense. If agents 'disagree' initially (Section 4.2), the outcome is not decidable in advance, but converges to some subset of the range (0,1). Consequently, the outcome may favour one or other of the opposing agent *Groups* more than the other, or possibly neither. In one sense the outcome is therefore highly unsatisfactory because there will always be one aggrieved party. Yet the existence of some outcome rather than none can be considered an advance. In some cases any tolerable outcome is highly desirable, such as peace negotiations that lead to an end to physical violence. An example is the eventual resolution of the Irish "troubles" of the latter half of the 20th century with the Good Friday agreement in April 1998. The possibility still exists that no consensus will be reached, and some simulations show that. In practical terms, it can be argued that consensus *must* emerge eventually, since continued conflict eventually becomes intolerable to people involved. The Arab-Israel dispute is an example of a long-running dispute which appears to have been intractable since the 1917 Balfour Declaration, which called for the establishment

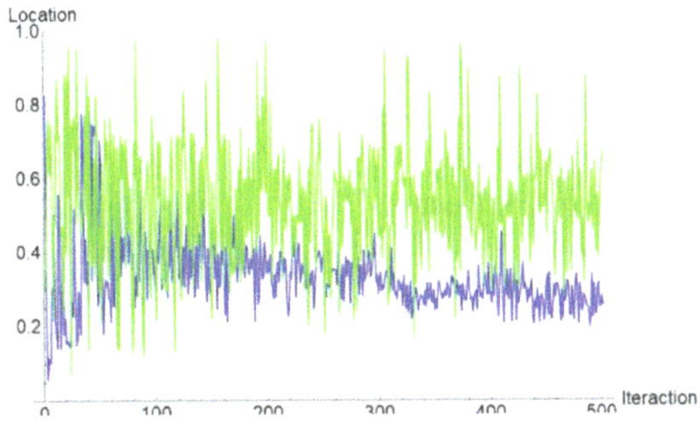

Figure 5: Random Interactions: *Group* size 10 (blue trace) and 100 (green trace)

of a homeland for the Jews in Palestine. Even that dispute has had its quieter moments (i.e. a tacit consensus), although conflict continues to flare up.

The conclusion that a single agent can have a fast significant effect on undecided agents (Sections 4.3) is noteworthy in the context of advertising or political campaigns. Arguably, influential figures were a major deciding factor in the 2016 Brexit referendum. However, the simulations of Section 4.4 show that an influential agent is unlikely to change the opinion of agents who are directly and strongly opposed in opinion. Section 4.5 illustrates a result that makes any prospect of consensus difficult: the more agents there are, the harder it is to agree.

As a suggestion for further work, we propose a change to deal with an unsatisfactory situation concerning the adequacy of the description of a *Group* comprising two distinct partitions, each with opposing opinions. In Section 3.3, the location of such a *Group*, or any other *Group*, is given as a weighted average of the locations of the members of the *Group*. This is problematic because it effectively describes an agent whose location is midway between two extremes: a situation that does not exist. A possible way out is to adopt the concept of a superposition of two opposing locations, which is essentially a quantum approach. The *Group Location* of a pair of agents would then be simultaneously two values, to be resolved on observation.

REFERENCES

[1] Rzevski, G. & Skobelev, P., *Managing Complexity*. WIT Press: Boston, 2014.

[2] Rzevski, G., A practical methodology for managing complexity. *Emergence: Complexity & Organization,* **13**(1/2), pp. 38–56, 2011.

[3] Epstein, J.M., Agent-based computational models and generative social science. *Complexity,* **4**(5), pp. 41–60. 1999. https://doi.org/10.1002/(sici)1099-0526(199905/06)4:5<41::aid-cplx9>3.3.co;2-6

[4] Epstein, J.M., Modeling civil violence: an agent-based computational approach. *Proceedings of the National Academy of Sciences of the USA*, **99**, pp. 7243–7250, 2002. https://doi.org/10.1073/pnas.092080199

[5] Lemos, C., Coelho, H. & Lopes, R.J., Agent-based modeling of social conflict, civil violence and revolution: state-of-the-art-Review and further prospects. *European Conference on Multi-Agent Systems (EUMAS),* pp. 124–138, 2013.

[6] Granovetter, M., Threshold models of collective behavior. *American Journal of Sociology,* **3**(6), pp. 1420–1443, 1978. https://doi.org/10.1086/226707

[7] Doran, J. Iruba: An Agent-Based Model of the Guerrilla War Process. *Proceedings 3rd Conference of the European Social Simulation Association, Koblenz*, pp. 198–205, 2005.

[8] Kim, J.W. & Hanneman, R.A., A computational model of worker protest. *Journal of Artificial Societies and Social Simulation,* **14(3)**, 2011. https://doi.org/10.18564/jasss.1770

[9] Gurr, T.R., *Why Men Rebel*, Princeton, 1970.

[10] Jones, A.J., *Game Theory: Mathematical Models of Conflict.* Horwood Publishing: West Sussex, 2000.

[11] Axtell, R.L., Epstein, J.M. & Young, H.P., The Emergence of Classes in a Multi-Agent Bargaining Model. *Social Dynamics*, eds. S.N. Durlauf & H.P. Young), MIT Press: West Sussex, 2001.

[10] Weisbuch, G., *Persistence of discrimination: Revisiting Axtell*, Epstein and Young, available at https://arxiv.org/pdf/1706.02573.pdf, 2017
[11] Liggett, T.M., *Interacting Particle Systems*, Springer, 1985
[12] Axelrod, R., Dissemination of culture: a model with local convergence and global polarization. *Journal of Conflict Resolution,* **41**, pp. 203–226, 1997. https://doi.org/10.1177/0022002797041002001
[13] Miguel, M.S., Eguiluz, V.M., Toral, R. & Klemm, K., Binary and multivariate stochastic models of consensus formation. *Computing in Science & Engineering,* **7**(5), pp. 67–73, 2005. https://doi.org/10.1109/mcse.2005.114
[14] Burt, G. *Conflict, Complexity and Mathematical Social Science*, Vol. 15, Emerald Group, 2010.
[15] Namatame, A. & Chen, S-H., *Agent-based Modelling and Network Dynamics*. Oxford University Press: Oxford, 2016.

APPENDIX A: BOUNDS ON GROUP INTERACTIONS OF AGENTS

This result is first mentioned in section 3.3. The resultant of the interaction between two agents X and Y, $Z = < X, Y >$, has location L_Z bounded by (the notation is explained below):

$$max\left(0, \frac{a_X\left(1-\frac{sa_X}{10}\right)}{a_Y\left(1+\frac{sa_Y}{10}\right)+b_Y\left(1+\frac{sb_Y}{10}\right)}\right) < L_Z < min\left(1, \frac{a_Y\left(1+\frac{sa_Y}{10}\right)}{a_X\left(1-\frac{sa_X}{10}\right)+b_X\left(1-\frac{sb_X}{10}\right)}\right).$$

Further, this result can be extended to multiple interactions between agents.

This result is proved by first considering two agents. Let agents X and Y have respective locations $\{L_X, L_Y\}$, and let their Beta distribution parameters be $\{a_X, b_X\}$ and $\{a_Y, b_Y\}$ respectively. Consider the first two steps of *Algorithm 1*. If $I \in (0,1)$ and $R \in (0,1)$ are factors representing influence and resistance respectively (see Section 3.2), and a' is an intermediate parameter, then, supposing $a_X < a_Y$ (without loss of generality):

$a' = Ia_X + (1-I)a_y$ (*Algorithm 1*, step 1), and

$a_Z = Ra_X + (1-R)a'$ (*Algorithm 1*, step 2).

$$\therefore a_Z = (I + R - IR)a_X + (1-(I+R-IR))a_Y$$

$$\therefore a_X < a_Z < a_Y \text{ and similarly } b_X < b_Z < b_Y. \quad \text{(A1)}$$

$$\therefore a_X + b_X < a_Z + b_Z < a_Y + b_Y. \quad \text{(A2)}$$

From (A2 and A1),

$$\left(\frac{a_X}{a_Y + b_Y}\right)\left(\frac{a_X + b_X}{a_Y + b_Y}\right) < \frac{a_Z}{a_Z + b_Z} < \left(\frac{a_Y}{a_Y + b_Y}\right)\left(\frac{a_Y + b_Y}{a_X + b_X}\right)$$

$$\therefore L_X\left(\frac{a_X+b_X}{a_Y+b_Y}\right) < L_Z < L_Y\left(\frac{a_Y+b_Y}{a_X+b_X}\right). \tag{A3}$$

Note that if $a_X+b_X > a_Y+b_Y$, equation (A3) reduces to $L_X < L_Z < L_Y$. In the subsequent perturbation stage, the limits set in (A3) are extended outwards accordingly. The standard error s corresponds to a given confidence level on a Normal density $N\left(0,\frac{c}{10}\right)$, where c is a placeholder for a_X, b_X, a_Y or b_Y. For example, at 95% confidence, the 2-tailed value of s is 1.96. The upper limit in (A3) is then *increased* to a 95% confidence maximum:

$$L_Y\left(\frac{a_Y+b_Y}{a_X+b_X}\right) \rightarrow \frac{a_Y\left(1+\frac{sa_Y}{10}\right)}{a_X\left(1-\frac{sa_X}{10}\right)+b_X\left(1-\frac{sb_X}{10}\right)}. \tag{A4}$$

Similarly, the lower limit in (A3) is then *decreased* to a 95% confidence minimum:

$$L_X\left(\frac{a_X+b_X}{a_Y+b_Y}\right) \rightarrow \frac{a_X\left(1-\frac{sa_X}{10}\right)}{a_Y\left(1+\frac{sa_Y}{10}\right)+b_Y\left(1+\frac{sb_Y}{10}\right)}. \tag{A5}$$

Lastly, step 4 of *Algorithm 1* limits the resulting confidence interval to (0,1). Therefore the final bound for L_Z is given by (A6). Inequality (A7) is an abbreviated version of (A6) in which M_L and M_U are written in place of the quotient arguments to the *max* and *min* functions.

$$max\left(0,\frac{a_X\left(1-\frac{sa_X}{10}\right)}{a_Y\left(1+\frac{sa_Y}{10}\right)+b_Y\left(1+\frac{sb_Y}{10}\right)}\right) < L_Z < min\left(1,\frac{a_Y\left(1+\frac{sa_Y}{10}\right)}{a_X\left(1-\frac{sa_X}{10}\right)+b_X\left(1-\frac{sb_X}{10}\right)}\right) \tag{A6}$$

$$max\left(0,M_L\left(X,Y\right)\right) < L_Z < min\left(1,M_U\left(X,Y\right)\right) \tag{A7}$$

Note that if agents X and Y have closely aligned sentiments (i.e. $L_X \sim L_Y$), the upper limit in (A6) is marginally greater than L_Y and the lower limit in (A6) is marginally less than L_X. The result (A6) applies generally for the interaction between any pair of agents. With each interaction the bounds are extended further. For n agents $\boldsymbol{X}=\{X_i\}_{i=1..n}$, and r interactions, let the set $\boldsymbol{X}_{[n,r]}$ be the resultant of all those interactions. Further let $\overline{X}_{[n,r]}$ be the weighted mean (Section 3.3) location of the members of $\boldsymbol{X}_{[n,r]}$ Therefore we can write (in the notation of (A7)):

$$max\left(0, min_r\left(M_L\left(\bar{X}_{[n,r]}\right)\right)\right) < L_Z < min\left(1, max_r\left(M_U\left(\bar{X}_{[n,r]}\right)\right)\right) \quad \text{(A8)}$$

Equation (A8) is the desired result for a *Group* of agents.

APPENDIX B: CONVERGENCE OF THE GROUP LOCATION

This result is first mentioned in section 3.3, and is an extension of the result of Appendix A. Convergence of the *Group Location* may be summarised by the statement:

*The interval in which the expected location of the resultant of an interaction between any two agents in a Group **X** is a subset of the corresponding interval for the previous interaction.*

This result shows that the expected *Group Location* converges to somewhere in (0,1) but does not specify where. The same notation that was used in Appendix A is used here. Let U_r be the upper limit of the expected value of the locations of members of the *Group **X*** after r interactions. Let V_r be the corresponding lower limit. Then

$$U_r = max_r\left(E\left(\boldsymbol{X}_{[n,r]}\right)\right);\ V_r = min_r\left(E\left(\boldsymbol{X}_{[n,r]}\right)\right) \quad \text{(B1)}$$

Let $L_{X_i}(r)$ be the location of agent X_i after r interactions. At interaction $r+1$, the location of X_i changes under the influence of an agent Y_i. Using $\Phi(0,s^2)$ for a single draw from a Normal distribution as in Section 3.2, for some $\mu_r \in (0,1)$:

$$L_{X_i}(r+1) = \mu_r L_{X_i}(r) + (1-\mu_r) L_{Y_i}(r) + \Phi(0,s^2) \quad \text{(B2)}$$

$$\therefore E\left(L_{X_i}(r+1)\right) = \mu_r E\left(L_{X_i}(r)\right) + (1-\mu_r) E\left(L_{Y_i}(r)\right) + 0$$

$$\therefore E\left(L_{X_i}(r+1)\right) \in (V_r, U_r) \in (0,1)$$

$$\therefore (V_{r+1}, U_{r+1}) \in (V_r, U_r) \quad \text{(B3)}$$

Equation (B3) establishes local convergence, effectively an expected reversion to the centre of the interval (0,1). The expected diminution of the interval in (B3) shows an expected progressive reduction in the variance of the location of *Group **X*** with increasing r.

ADAPTIVE DESIGN: WATER CURTAINS FOR WAYOUT FINDING IN HUB SPACES

F.A. PONZIANI[1,2], A. TINABURRI[1,2], & V. RICCI[1]
[1]Central Direction for prevention and technical safety – Department of firefighters, of the public rescue and civil defence – Ministry of the Interior – Italy.
[2]University of Rome Tor Vergata – Italy.

ABSTRACT

The aim of this study is to explore some features of a complex system arising from the interactions of a fire stream in a hub space layout, with fire protection through water curtains issued by edge nozzles activated by smoke detectors. The hub layout represents the landside part of an airport terminal, made of clusters of semi-enclosed isles open to the inter-connected enclosed spaces that form a series of longitudinal paths with services and utilities. Once a fire source emits matter (smoke) and energy (enthalpy) out of one of the isle, in the absence of any fire protection barrier, the stream wanders following its buoyancy and the boundary conditions filling the available spaces inside the hub, making the occupants' conditions untenable. The design of water curtains that are activated in response to the fire onset may help to limit the dangerous spread of the fire stream and to support in the unfolding of protected paths for egress. While the activation of the water curtains in the proximity of the fire source once a threshold value is reached is the classical approach, a different design strategy is here investigated with CFD modelling based on a sequence of adaptive responses of the hub layout to the fire stream.
Keywords: adaptation, CFD, complex systems, edge nozzle, fire, FSE, water curtain.

1 INTRODUCTION: FIRE AND CROWD ECOSYSTEMS

This study is part of the research programme meant to explore Crowd and Fire Dynamics in their essence of Complex Systems [1]. In these kind of systems, the relationships between parts give meaning to the entire system, thus showing an Ecosystem. This is the case for a fire developing in an environment, where the interactions between fire and environment are responsible for the dynamic that will emerge, and when people get involved adding more complexity to the system, the interactions will include people behaviour that will bring about another kind of dynamic of evolution. While fire and crowd dynamics deserve understanding and studying on their own, so that each one of those two complex systems may reveal their nature independently – that matters anyway when there is no coupling between fire and crowd or when a design of a space layout is a task that requires knowledge of fire dynamics principles or of crowd dynamics principles alone – one more added value is understanding and studying of fire and crowd dynamics together – that matters when the coupling of fire and crowd streams is a major topic in design and management of the space layout [2].

Be fire and people independent, or not, this study presents an innovative viewpoint: fire and people form an ecosystem where the typical characteristics of fire dynamics (i.e. smoke streams) and of crowd dynamics (i.e. people streams) interact with the environment and evolve with emerging properties depending on the interactions. The shift of the focus is from the single parts to the interactions [3]; for smoke management: stopping and channelling of smoke stream with interactive means such as vents and curtains that activate with a sequence linked to the smoke advancement; for egress management: way-out finding activities linked to paths pre-defined or unfolding depending on the advancement of the dangerous menace. Interactions are one typical aspect of Complex Systems from which emergent behaviour may appear in several forms, unpredictable yet not random [4]: the understanding of what kind of

DOI: 10.2495/DNE-V13-N3-294-306

emerging patterns arise is one main problem of this research project, where the fire stream may evolve with patterns in different ways depending just on the interactions between the system components, such as smoke advancing front and active protection devices, and environment layout.

This study deals with a hub design with water curtains that activate in sequence to help reduce the effects of smoke spreading from a fire source. The hub layout represents the landside part of an airport terminal, made of clusters of semi-enclosed isles open to the inter-connected enclosed spaces that form a series of longitudinal paths with services and utilities.

Once a fire source emits its stream of matter (smoke) and energy (enthalpy) out of one of the isle, in the absence of any fire protection barrier, the stream wanders following its buoyancy and the boundary conditions filling the available spaces inside the hub, sooner or later making the occupants' conditions untenable.

Water curtains that activate in response to the fire onset may help to limit the dangerous consequences of the spread of the fire stream – at least for some time useful to start control or suppression of the fire source or to start egress of occupants – and to support in the unfolding of protected paths for way-out – thus making adaptation of the system a focus for egress – especially when supported by some means for limiting fire growth at the source.

Instead of activating the water curtains only in the proximity of the fire source once a threshold value is reached, as in a standard design, the activation of the water curtains is based on a sequence of variable responses of the hub layout to the fire stream. At the advancement of the fire/smoke stream, while the water curtains activating near the fire source may help in controlling and reducing the spread, other water curtains along the hub are activated if critical targets – not necessarily located near the edge nozzles – reach a threshold value due to the exposure to the stream advancement. Such exposure, that may be of temperature, heat flux, visibility at the targets, then triggers the dynamical response of the fire protection system, thus letting the system adapt to the fire stream evolution. CFD modelling allows to study the interactions of the fire/smoke stream with the hub layout and the response of the water curtains and the protected paths unfolding for way-out. In this way, the network made of active protection devices (smoke detectors and water curtains linked for activation) delivers a connection through the system that offers, by means of decentralized nodes (the devices themselves), the possibility for the system to adapt to the evolution dynamics occurring (i.e. the fire/smoke current) and deploy a resilient response (i.e. avoiding a full fire/smoke involvement in – at least some part of – the environment), thus enabling the formation of protected spaces (i.e. paths for way-out for people inside).

2 FIRE ENGINEERING DESIGN: SPACE LAYOUT AND CFD

The space layout of the hub considers an airport that is composed of one base floor level and a single height ceiling, representing the landside space of the terminal, with an overall strip-like shape in plan. On the left side, two partially open square spaces connect to a central concourse, while on the slightly narrower right side a couple of linear walls with turning end corners mark the line. There are four doors to the outside, two in the front part of the concourse and one for each end side, while there are four passageways for each square space isle that partially open to the concourse and two wide passageways that merge the linear walls and the concourse. The total length of the airport space is 60 m, 32 m on the left and 28 m on the right, the total width is 12 m on the left and 8 m on the right, the total height is 4 m.

The two square-sided isles on the left have a ceiling on top at the height of 3 m, while there is always an open space on top of the floor bounded by the linear walls up to the ceiling at the height of 4 m that is running all over the airport. The dimension of the doors, passageways and exits – essentially the width – can be varied based on the code followed for the design, where several dimensions have been considered in this study, with one group of them here reported in the remainder.

The following Figs 1 and 2 show a plan view and a rendering of the airport.

The following Table 1 summarizes the zones into which the hub space is divided for functional identification.

Once defined the hub layout, a fire engineering design by means of a CFD study is required to examine the impact of the layout on the development of a fire starting from a spot in one stand. Due to the fire safety provisions requiring few to no combustible materials in such

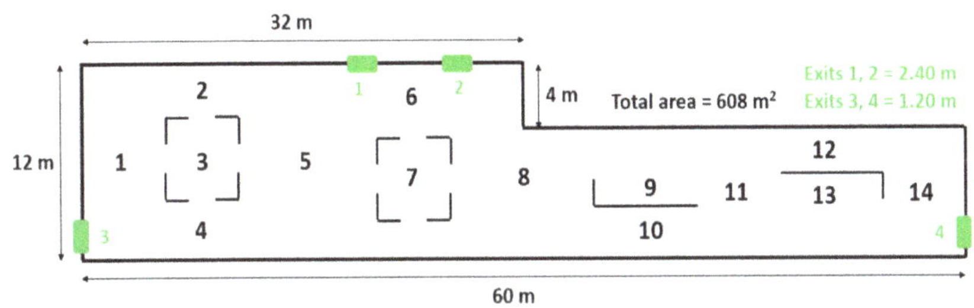

Figure 1: Plan view of the Hub.

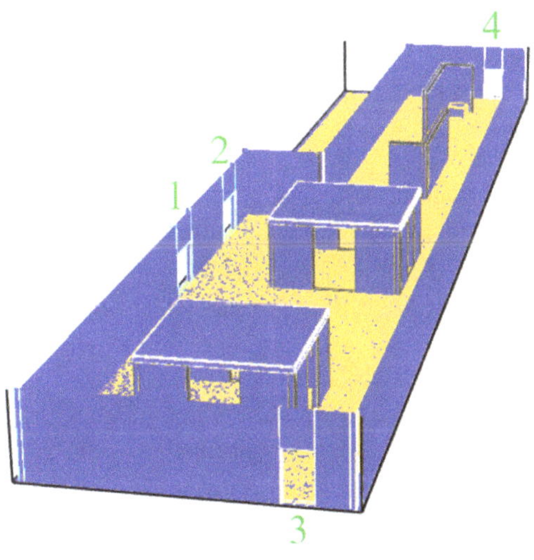

Figure 2: The Hub space.

Table 1: Functional identification of zones in the Airport hub.

Zone	*Identification*	*Function*	*Ceiling height*
1–5–8–11–14	Central concourse	Capacity for gathering people/smoke	3.80 m
2–4–6–10–12	Sideway concourse	Capacity for moving people/smoke	3.80 m
3–7	Square-sided isles	Space for stand	2.80 m
9–13	Linear walls	Space for stand	3.80 m

critical parts in the concourse needed to guide the occupants along way-out finding, the only places where a fire could be started – accidentally or voluntarily – are in the stands, albeit with a minor firepower allowed (i.e. by means of proper items with low heat/smoke production rates) in order to avoid a fast spread from the ignited source.

The CFD package Fire Dynamic Simulator (FDS) and its companion visualizer Smokeview (versions 6) are used in LES mode for modelling turbulence [5]. One domain with single mesh discretization is set-up for modelling the hub space; inert materials are assigned to describe the boundaries (i.e. floor, walls, ceilings) and the fire spot is modelled as a burner emitting an assigned firepower, where the reaction type is for plastic material combustion. Two scenarios are considered for the position of the fire source: in a corner in the central isle in the left part of the airport (F1) and in the corner of the right wall in the right end side of the airport (F2). This choice makes it possible to examine the spreading of the smoke stream moving along the hub from two opposite sides.

The CFD domain reproduces the hub layout shown in Fig. 2, with the following Table 2 summarizing the main input properties.

Once defined the model, CFD is used to study the interactions arising between the fire and the environment: the evolution of the dynamics modelled will be observed in such a way to start with a base case without active protection systems aimed at controlling the fire spread, followed by the introduction of active protection measures inserted to bias the dynamics of the fire. The active protection system considered in this study combines water curtains, placed in critical parts of the layout, with smoke detectors used for their activation.

Depending on the advancement of the fire/smoke stream, the smoke detectors will trigger the water curtains according to the design strategy, either following (adaptive design) or anticipating (generative design) the effects of the fire. The design key parameters to be assessed during the fire evolution may be the visibility loss, the gas temperature rise, the flow field vorticity induced by the water curtains interacting with the smoke. This methodology approach is thought and applied to try to let the system organize by itself with action-reaction behaviours biased on the interactions appearing during the very evolution dynamics occurring (i.e. different evolution dynamics lead to different interactions, hence to different action-reaction behaviours, hence to different organization of the system).

3 DESIGN BASIS

The first design step is modelling the fire evolution in the hub without water curtains operating. Among all the possible scenarios, two key alternative events are chosen: a fire starting in

Table 2: Input properties for CFD modelling.

Property	*Setup*
Domain	Single mesh-grid
Mesh-grid	Cubic cells
Turbulence modelling	Large Eddy Simulation (LES) with Sub Grid Scale (SGS)
SGS turbulence	Deardorff
Fire source	Burner
Firepower	Constant release
Reaction type	Plastic combustion
Soot emission	Constant yield

a square-sided isle on the left part of the hub (zone 7, scenario F1) and a fire starting at a guide wall on the right part of the hub (zone 13, scenario F2).

The following Fig. 3 shows the fire source locations, that are chosen since they host stands where combustible installations and appliances may temporarily be present. The presence/absence of a ceiling above the fire source lower than that of the hub is also taken into account: the stand in the isle has its own incombustible ceiling (at a lower height of 3 m), while the guide wall has a height limited to 3 m and the space above it is open up to the hub roof level (being 4 m throughout the terminal space). So, two differing kind of fire/smoke stream originate from a fire source of the same power and type.

The following Tables 3 and 4 summarize the fire and environmental parameters adopted for CFD modelling.

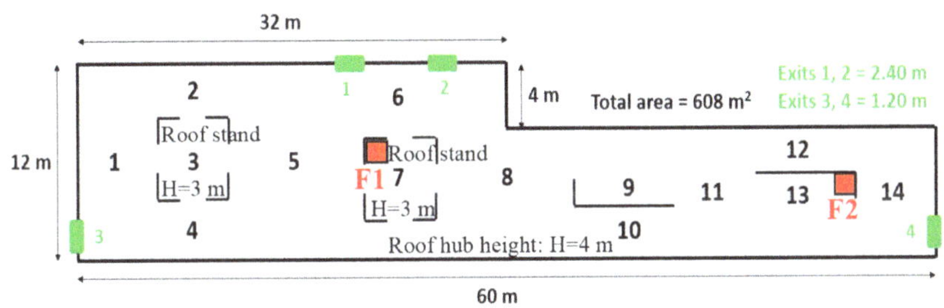

Figure 3: Alternative fire scenarios.

Table 3: Fire parameters for CFD modelling.

Parameter	*Value*
Domain	60 m length 12 m width 4 m height
Mesh-grid	Cubic cells. Reference size: 20 cm. Sensitivity analysis: minimum size 10 cm
Fire source	Corner in the left central square isle or in the right guide wall
Firepower	Steady-state 100 kW
Combustion reaction	Polyurethane
Soot yield	0.10
Spot with fire	Stand in isle or wall
Above top of fire	Lower own ceiling for isle or higher hub ceiling for wall stands
Smoke detectors	At ceiling when active
Water curtains	At ceiling when active

Table 4: Environmental parameters for CFD modelling.

Parameter	*Value*
Temperature	288 K
Density	1.22 kg m^{-3}
Specific heat at constant pressure	1.0 kJ kg^{-1} K^{-1}
Gravity field	9.81 m s^{-2}

The second design step assesses the effectiveness of the water curtains for the selected fire scenarios. Each curtain is formed by a series of nozzles placed aside at a same properly chosen height, thus forming a horizontal line, with an opening slot issuing a downward water flow forming a fan shaped curtain opening downward too. The set of sided single fan shaped curtains forms the overall water curtain as a whole. While not explicitly considered because not necessary for this study, it is assumed that a water supply is available to guarantee the required flow rate at the open nozzles (differing from sprinkler heads that are closed by a thermo-sensible element). Experimental testing is almost always necessary to verify and confirm the hydraulic design assumptions; some parameters adopted in this study to model the water flow come from a previous work carried out by the authors [6].

The water curtains are placed in some critical parts of the hub, i.e. in the central concourse and in the stand zones; their activation strategy is one major topic dealt with in this study. When there is no automatic control of the activation, the water curtains discharge their flow at the starting of the hydraulic system that, at most, can be triggered by a manual command scheduled at the discovery of the fire onset: we refer to this strategy as to the standard design. When there is an automatic control of activation (i.e. by smoke detectors), the water discharge is initiated after a logic sequence is performed: this means that the hydraulic system operation is linked to the control strategy.

When the activation of the water curtains by the controllers follows the event (i.e. the smoke advancement from the fire source), we refer to this as to the *adaptive* design. When the activation of the water curtains by the controller anticipates the event, we refer to this as to the *generative* design.

It is worth noting that the connection between the hydraulic system and the mastering controllers need not be necessarily localized near the water curtains, as it would be in the case of a sprinkler system, but it can be designed choosing the controller devices position in the layout in such a way to delocalize, as it would be in the case of a pre-action sprinkler system. This allows the use of multiple control devices thus enhancing the possibility of early detection and counter-measure of the fire threat. In that way, it is possible to guide the active protection along the very fire dynamic that is evolving in the system, either in defence (adaptive) or in attack (generative). It could be noted that this approach resembles closely what is done in the process control field, where the adaptive strategy is known as feed-back control mode and the generative strategy is called feed-forward control mode [7].

The following Table 5 summarizes the water curtains hydraulic and control logic parameters adopted for CFD modelling.

The following Tables 6 and 7 summarize respectively the location of the water curtains and of the smoke detectors and the sequences of activation, defining the type of design.

The protection devices can be classified depending on their function into detectors and activable ones. A strategy matrix can thus be designed where the columns are the detectors (in our study smoke sensors S_j) and the rows are the activable devices (in our study water curtains W_i), whose elements are assigned a value 0 or 1 depending on their link. A logic code could then be developed combining two data sets: the first part identifies the fire scenario and the following the strategy matrix per row. This code defines the response of the whole system along its evolution dynamics [8], identifying the design strategy that might be implemented on a control panel. In that way the system self-organization is made evident, with the fire protection system autonomously able to respond to a fire/smoke event with the decentralized, and not centrally controlled, protection devices linked for activation.

Table 5: Water curtains hydraulic and control logic parameters for CFD modelling.

Parameter	*Value*
Nominal flow rate	18 l min^{-1}
Frontal screen opening angle	120°
Side screen opening angle	6°
Pressure at nozzle	6 barg
Initial velocity	3.75 m s^{-1}
Mean droplet diameter	200 μm
Statistical droplet distribution	Rosin-Rammler-Lognormal (default FDS)
Injection rate	5000 particles per second (default FDS)
Controllers of activation	Smoke detectors activated with the single parameter Heskestad model (default FDS)

Table 6: Active protection devices: water curtains and smoke detectors.

Device	*Identification*	*Location*	*Zone*	*Height of device*
Water curtains at concourse	W1	Central concourse	8	3.80 m
Water curtains at concourse	W4	Central concourse	11	2.80 m
Water curtains at stand isle	W2	Square sided-isle	3	2.80 m
Water curtains at stand isle	W3	Square sided-isle	7	2.80 m
Water curtains at stand isle	W4	Linear wall	9	2.80 m
Water curtains at stand isle	W4	Linear wall	13	2.80 m
Smoke detector at concourse	S5	Central concourse	5	3.60 m
Smoke detector at concourse	S1	Central concourse	8	3.60 m
Smoke detector at concourse	S4	Central concourse	11	3.60 m
Smoke detector at stand isle	S2	Square sided-isle	3	2.80 m
Smoke detector at stand isle	S3	Square sided-isle	7	2.80 m

Table 7: Active protection devices: water curtains and smoke detectors.

Fire scenario	*Smoke detection*	*Water curtains activation*	*Design type*
Stand in the isle 7	S3	W3	Adaptive
(scenario F1)	S1	W1	
	S2	W2	
	S4	W4	
Stand near the wall in zone 13	S4	W4	Generative
(scenario F2)	S4	W1	
	S1	W3	
	S5	W2	

Table 8: Evolution response: system code strings

Fire scenario	W_iS_j : 1st row	W_iS_j : 2nd row	W_iS_j : 3rd row	W_iS_j : 4th row
(F1) 0 1	1 0 0 0 0	0 1 0 0 0	0 0 1 0 0	0 0 0 1 0
System code string: 0110000010000010000010 (*Adaptive design*)				
(F2) 1 0	0 0 0 1 0	0 0 0 0 1	1 0 0 0 0	0 0 0 1 0
System code string: 1000010000011000000010 (*Generative design*)				

In our study two fire scenarios F1 and F2 are considered, with the assigned strings 01 and 10, with four water curtains and five smoke detectors. The preceding Table 8 shows the system code strings thus obtained.

4 CFD: BASELINE DESIGN WITHOUT ACTIVE PROTECTION SYSTEMS

The baseline design case is chosen with the fire evolving without any water curtains operating. Fire scenario F1 is the base case for the analysis of the effect of the adaptive strategy while fire scenario F2 is the base case for the generative strategy. Some selected results of the CFD simulations are shown in the following Figs 4 and 5.

The fire dynamics in the absence of active protection intervention is simulated with a 20 cm mesh grid for a time interval of 120 s (ignition at time 0), long enough to allow the comparison with the simulation where the water curtains are triggered by the detectors in response to the smoke advancement.

In fire safety engineering, when dealing with the early stage of a fire, the main threat for the occupants is represented by the smoke. Without any active protection system, in the layout studied the smoke spreads throughout the whole compartment mainly flowing at the

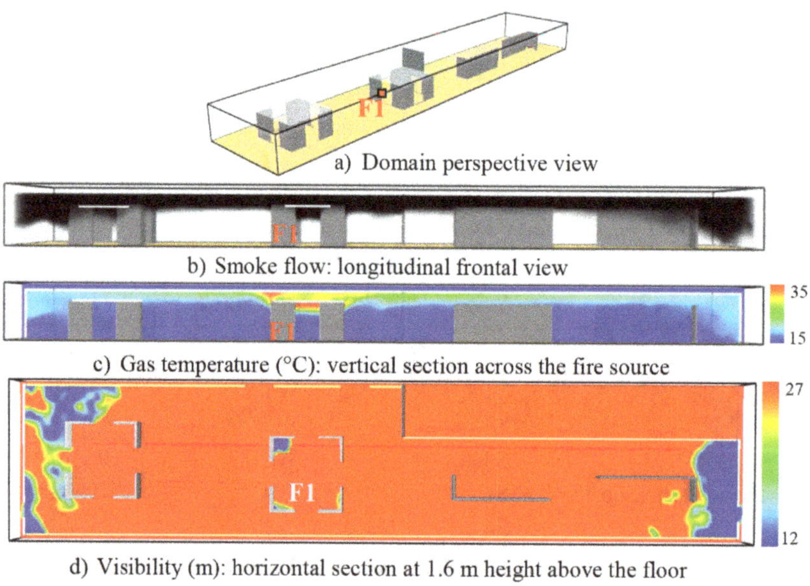

Figure 4: Scenario F1 without water curtains operating – 120 s from ignition.

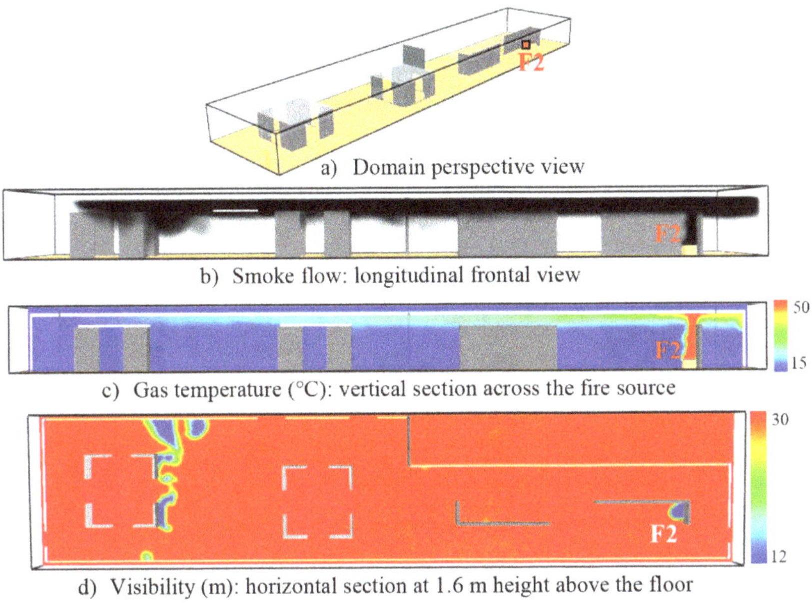

Figure 5: Scenario F2 without water curtains operating – 120 s from ignition.

ceiling level and reaches the side boundaries in a time span in the order of a few minutes, even with a low intensity fire source. It is worth noting the increased vorticity appearing in correspondence of the interaction of the smoke layer with obstacles far away from the fire source and the shape of the smoke front while advancing freely along the ceiling.

5 CFD: ACTIVE PROTECTION ADAPTIVE DESIGN

In the adaptive design the smoke detector activates its linked water curtains in a sequence that follows the smoke front advancement. The following Figs 6 and 7 show some selected CFD results and the domain for the 10 cm mesh with the smoke detectors and water curtains: the activated smoke detectors and the linked water curtains are shown respectively in red and blue colours. The smoke detector S3 is the first to trigger the water curtains W3 after ~ 6 s; S1 triggers W1 after ~24 s (~26 s for the coarsest grid). Water flow rendering is made visible only in the first two snapshots (Fig. 6 a and b) for the sake of clarity.

At 120 s, when the simulation is stopped, there is no further smoke detector activation. The water curtains W3 favour the smoke accumulation inside the stand on fire (recall that there is no direct intervention on the fire source, which is assumed steady-state). The water curtain W1 obstacles the smoke propagation beyond it; the smoke outflowing from the stand on fire is therefore forced to direct mainly on the left side of the hub space.

Compared with the baseline case simulation, water curtains induce zones of high vorticity at the floor level. While the smoke is still propagating mainly along the ceiling, smoke destratification is observed around the stand on fire and in the space between the stand and the water curtain W1, impacting negatively on the visibility conditions in this zone. With the low intensity fire source assumed, the gas temperature rise at ceiling level is in the order of 10°C, while in the baseline case is in the order of 20°C.

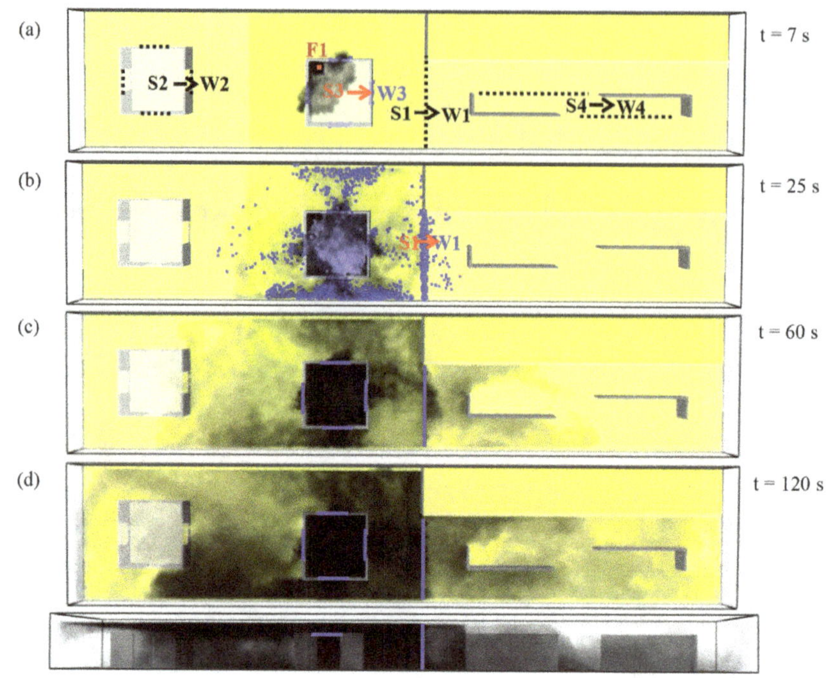

Figure 6: Scenario F1 with water curtains (*Adaptive design*) – Smoke flow sequence: top view at 7, 25, 60 and 120 s (with the hub ceiling clipped); longitudinal front view at 120 s.

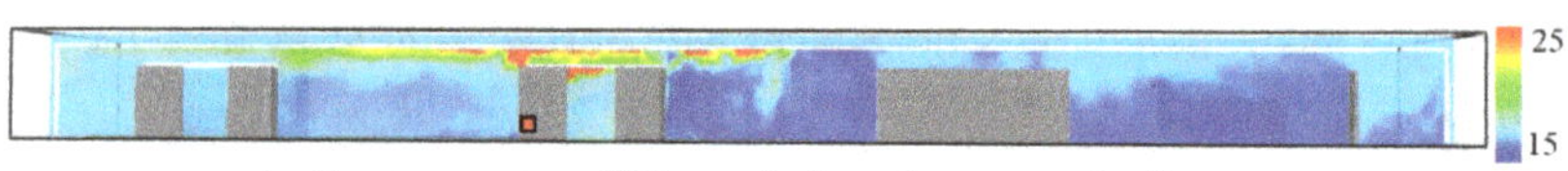

a) Gas temperature (°C): vertical section across the fire source

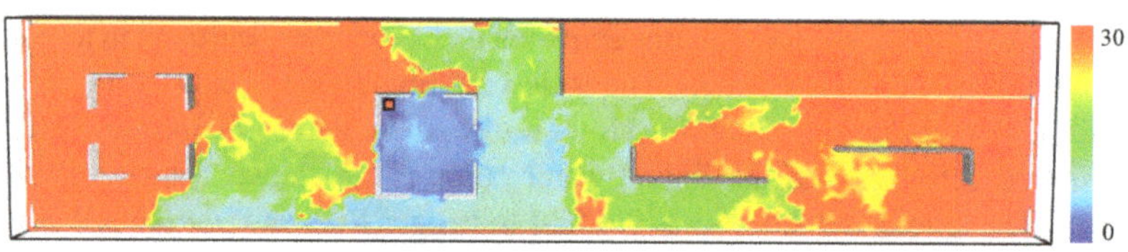

b) Visibility (m): horizontal section at 1.6 m height above the floor

Figure 7: Scenario F1 with water curtains (*Adaptive design*) – 120 s from ignition.

6 CFD: ACTIVE PROTECTION GENERATIVE DESIGN

In the generative design the smoke detector activates its linked water curtains in a sequence that anticipates the smoke front advancement. The following Figs 8 and 9 show some selected CFD results and the domain for the 10 cm mesh with the smoke detectors and water curtains: the activated smoke detectors and the linked water curtains are shown respectively in red and blue colours. The smoke detector S4 is the first to trigger the water curtains W4 and W1 after ~17 s (~22 s for the coarsest grid); S1 triggers W3 after ~90 s (~92 s for the coarsest grid). Water flow rendering is hidden for the sake of clarity.

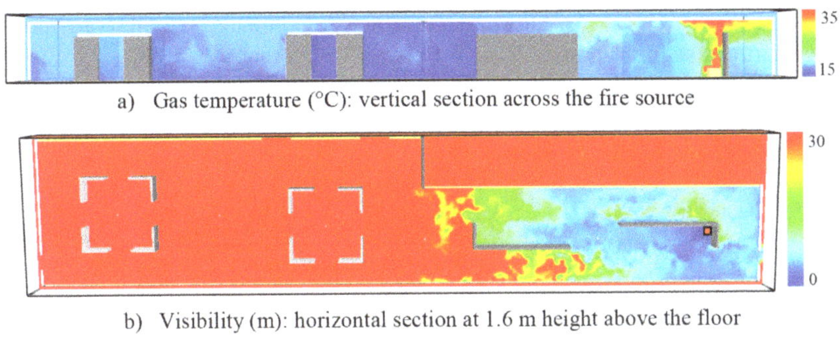

Figure 8: Scenario F2 with water curtains (*Generative design*) – 120 s from ignition.

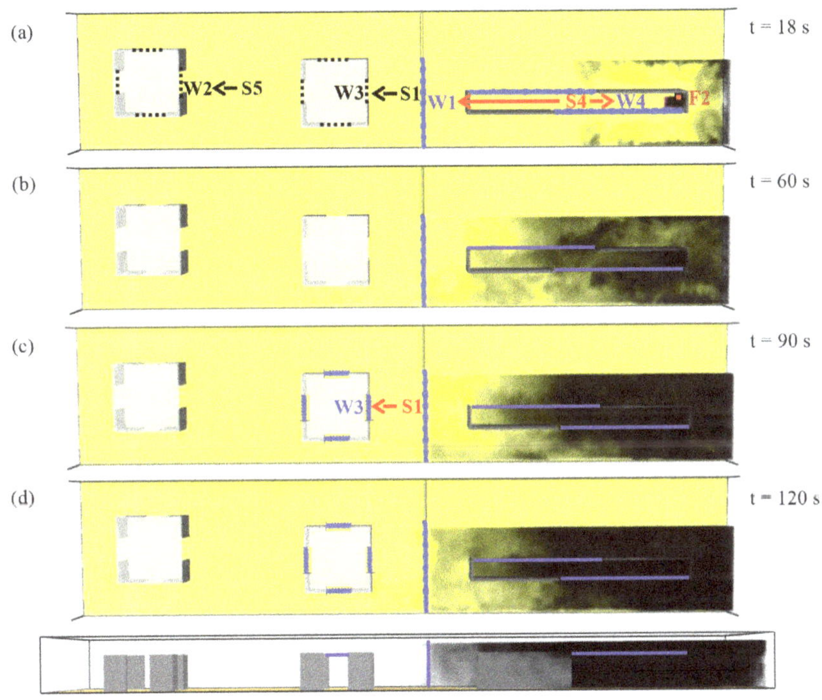

Figure 9: Scenario F2 with water curtains (*Generative design*) – Smoke flow sequence: top view at 18, 60, 90 and 120 s (with the hub ceiling clipped); longitudinal front view at 120 s.

At 120 s, when the simulation is stopped, there is no further smoke detector activation. The water curtains W4 are deliberately placed at a lower level than the hub ceiling so that the smoke produced by the fire can overpass them in order to activate further protective devices (generative design strategy meant to anticipate the active protection response). Since W4 completely surround the fire source filling the gaps between the linear walls, smoke accumulation in zones 11 and 13 is still observed. The water curtain W1 is activated simultaneously with W4 well in advance of the smoke front arrival. The smoke outflowing from the stand on fire directs mainly on the right side of the hub space, reaching a critical state after ~90 s when smoke detector S1 is activated. With a generative design, the nearest sensible target, i.e. the stand in the isle 7, is protected by the water curtain W3. Compared with the previous simulation, water curtains seem to induce less vorticity at the floor level. While the smoke is still propagating mainly along the ceiling, smoke destratification is observed around the stand on fire. The gas temperature rise at ceiling level is now in the order of 20 °C, while in the baseline case is in the order of 40 °C.

7 CONCLUSION: ECOSYSTEMS

The fire and crowd ecosystems, alone or coupled, live on interactions that define their evolution dynamics. Critical events may appear, disappear, mutually interfere, depending on their nature and the constraints from the environment. One such example of ecosystem has been exploited in this study, where active protection systems with water curtains activated by smoke detectors have been designed to respond to the advancement of the smoke stream produced by the fire source (i.e. the fire stream), thus providing a support for enhancing way-out finding (i.e. the crowd stream) either following the events (adaptive design: following the smoke advancement) or anticipating them (generative design: anticipating the smoke advancement). While further refinement, easily added, is needed for direct application, requiring evaluation of the relative strengths of the fire stream and of the water curtains with higher values of the parameters (i.e. firepower, flow rate), together with experimental tests for parameters assessment, the design process depicted here shows an innovative viewpoint in fire research and engineering that takes into account the very nature of ecosystem of such complex systems as fire and crowd systems. Coupling CFD modelling and experimental campaigns of water curtains may lead to efficient and improved solutions, where outcomes (i.e. reduced smoke spreading, thermal screen) and drawbacks (i.e. smoke crossing and rolling vortex) of the water curtains are balanced and active protection is enhanced by means of proper design (i.e. denser water screens, curtains in series) and of further interacting devices (i.e. smoke curtains, smoke channelling and venting) [9–12]. Furthermore, by recognising the complexity of systems involving fire and crowd streams development may help to deepen the understanding of those complex systems and to approach the related fire safety engineering with innovative and practical solutions and design strategies.

REFERENCES

[1] Erdi, P., *Complexity Explained*, Springer: Berlin, 2008.

[2] Hurley, M.J. (ed), *The SFPE Handbook of Fire Protection Engineering*, 5th Edition, Springer-Verlag: Berlin and New York, 2016.

[3] Haken, H., *Synergetics. Introduction and Advanced Topics*, Springer: Berlin, 2004.

[4] Rzevski, G., Harnessing the power of self-organization. *International Journal of Design & Nature and Ecodynamics*, **11**(4), pp. 483–494, 2016. https://doi.org/10.2495/dne-v11-n4-483-494

[5] Mc Grattan, K., Hostikka, S., Floyd, J., Weinschenk, C. & Overholt, K. (eds), *Fire Dynamics Simulator v.6. Technical Reference Guide*, NIST Special Publication 1018-1, 2015.

[6] Ponziani, F.A., Tinaburri, A. & Ricci, V., Water curtains for fire protection: experimental tests and CFD modelling. *International Journal of Safety and Security Engineering,* **8**(1), pp. 1–9, 2018. https://doi.org/10.2495/safe170091

[7] Shinskey, F.G., *Process Control Systems*, Mc Graw-Hill: New York, 1996.

[8] Holland, J.P., *Adaptation in Natural and Artificial Systems*, MIT Press: Cambridge MA, 1992.
[9] Roytman, M.Ya., *Principles of Fire Safety Standards for Building Construction*, Amerind: USA, 1975.
[10] Centro Studi Esperienze CNVVF (ed.), *Barriere Idriche*, Rome, 1999 (in Italian).
[11] Choi, C.L., Radiation blockage effects by water curtain. *International Journal on Engineering Performance-Based Fire Codes*, **6(4)**, pp. 248–254, 2004.
[12] Cheung, W.Y., Radiation Blockage of Water Curtains, *International Journal on Engineering Performance-Based Fire Codes*, **1**, pp. 7–13, 2009.

CONCEPTUAL DESIGN OF SMART FARMING SOLUTION FOR PRECISE AGRICULTURE

D. BUDAEV, A. LADA, E. SIMONOVA, P. SKOBELEV, V. TRAVIN, O.YALOVENKO, G. VOSHCHUK & A.ZHILYAEV
Knowledge Genesis Group, Smart Solutions, Samara State Technical University, Samara University, Moscow Timiryazev Agricultural Academy, Peschanokopskaya Agro Group.

ABSTRACT
The paper presents precise agriculture as a complex adaptive system with high level of uncertainty and dynamics, in which knowledge is forming experimentally and in a very enterprise-specific way. Is there any opportunity to learn the best practices from advanced precise farmers, transfer their knowledge and support everyday decision making for regular farmers? The concept of Smart Farming as an augmented AI solution for precise agriculture is proposed. The solution is designed as a digital eco-system (system of systems) of smart services, where each service, in its turn, is an autonomous AI system. The paper also discusses functionality of smart services for precise agriculture and the service-oriented architecture of the solution with p2p interaction of services. Ontology-driven knowledge base and multi-agent technology are considered as the key technologies of the solution. The virtual "round table" for coordinated decision making of smart services is introduced. Finally, the paper presents results of the first applications, as well as the future steps and expected results.
Keywords: multi-agent technology, ontology, precise agriculture, real time, smart farming.

1 INTRODUCTION

The growing human population generates the increasing demand for more efficient food production under conditions of climate changes, constrained land resources, etc. One of the new perspective technologies is *precise agriculture* [1–3], which assumes ongoing monitoring of crops and fast precise reaction to unpredicted events in real time.

The quality and efficiency of precise agriculture can be significantly improved by using smart systems with the following capabilities for real-time decision making:

- real-time satellite- and drone-based pattern recognition of problematic situations with the use of hyper spectrum cameras (illness, insects, etc.);
- real-time event-driven adaptive resource management for machines, pesticides and other resources, including unmanned vehicles.

These opportunities could be designed and implemented as smart web-services available for farmers via constantly running Internet portal and mobile applications [4].

In this paper, we propose to design a new Smart Farming solution as a digital eco-system of smart services with the use of knowledge base and multi-agent technology. In the proposed concept, precise agriculture is considered as a complex adaptive system with high level of uncertainty and dynamics. The domain knowledge of precise agriculture for open-air farming is usually formed experimentally and in a very enterprise-specific way. It is not repeatable and not transferred from one enterprise to another. Is there any opportunity to learn the best practices from advanced precise farmers, formalize and transfer their knowledge and support everyday decision making for regular farmers?

The concept of Smart Farming as an augmented Artificial Intelligence (AI) solution for precise agriculture is considered. The solution is designed as a knowledge-based digital

DOI: 10.2495/DNE-V13-N3-307-314

eco-system (system of systems) of smart services, where each service, in its turn, is considered as an autonomous AI system. The proposed concept will provide the opportunity to consider the decision-making process in agricultural enterprise as a negotiation process focused on solving conflicts and finding consensus among agents of experts in soil, wheat production, fertilizers, pesticides, finance, machines, etc.

Section 2 introduces the proposed concept and functionality of knowledge-based digital platform and eco-system for agriculture. The architecture of eco-system of smart services, as well as the virtual "round table" for coordinated decision making of agents representing smart services are presented in Section 3. Section 4 is focused on the first application and expected results. Section 5 presents the summary and outlines the next steps of developments.

2 THE FUNCTIONALITY OF SMART SERVICES FOR PRECISE FARMERS

According to Gartner's definition, "A digital ecosystem is an interdependent group of enterprises, people and/or things that share standardized digital platforms for a mutually beneficial purpose (such as commercial gain, innovation or common interest). Digital ecosystems enable you to interact with customers, partners, adjacent industries — even your competition" [5].

In our vision, a digital ecosystem is an open, knowledge-based, distributed and adaptive system of smart services ("system of systems"), which is able to self-organize and demonstrate other properties of complex systems [6]. As an example, let us consider how two key smart services could work autonomously and adaptively in strong cooperation:

- at any time smart pattern recognition can trigger re-scheduling of resources, for example, involving more fertilizers or pesticides;
- execution of plan may trigger new pattern recognition for control of results.

These two services can be considered as the "left" and "right" parts of agricultural "brain": one part is used for work with unstructured information, and the second part – for structural information, logical reasoning and calculations, helping farmers to reduce time-to-go.

In the proposed concept, smart services are considered as autonomous AI systems based on multi-agent technology – each service is represented by a software agent which is able to react to events, make decisions and coordinate them with other agents or humans [6]. The functionality of the solution includes the following agents of services (Fig. 1):

- Agents of Fields – multi-agent system for managing fields, including crop allocation, scheduling, optimizing and controlling;
- Agents of Monitoring – satellite / drone–based multi-agent system for recognizing patterns of illness, insects, etc.;
- Agents of Machines – multi-agent system for managing groups of machines for providing on-demand services for cultivating fields, bringing pesticides, etc.
- Agents of Pesticides – multi-agent system for bringing pesticides to agricultural fields;
- Agents of Fertilizers – multi-agent system for bringing fertilizers to fields;
- Agents of Finance – multi-agent system for finance planning, etc.

Agents of services can be implemented by the specific multi-agent system of low-level agents representing orders and resources, technology processes and tasks, products, spare parts and other entities of agricultural business.

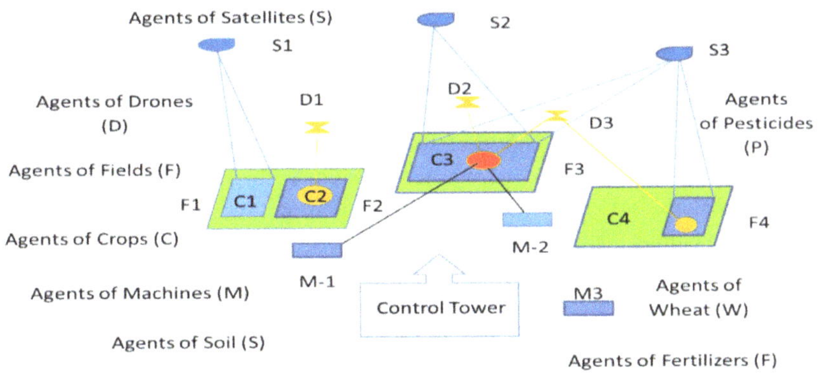

Figure 1: The multi-agent eco-system for supporting self-organization among services of satellites (S), drones (D), fields (F), machines (M) and others.

In the designed eco-system the images from satellites are continuously processed with calculation of NDVI indexes. They provide the global view on the fields, which helps recognize problematic areas for further, more detailed, investigation by swarms of drones or machines. When a problem pattern is recognized, the system recommends actions for solving the problem with the use of knowledge base, and makes adaptive re-scheduling of machines. Detected problematic zones require more attention from satellites and drones – until it becomes clear that the problem is solved.

3 THE ARCHITECTURE OF THE SOLUTION

The designed solution will be organized as service-oriented architecture with enterprise service bus and p2p communication of smart services presented in Fig. 2.

The main part of the solution will be the knowledge base of precise farmers, which contains the following components:

- agricultural problem domain ontology represented by a semantic network of classes of concepts and relations with specific parts for wheat, pesticides, etc.;
- ontological model of enterprise and its current scene with the states of all objects and processes, which will contain specific instances of the above-mentioned classes, for example, information on history and state of fields, machines and other resources.

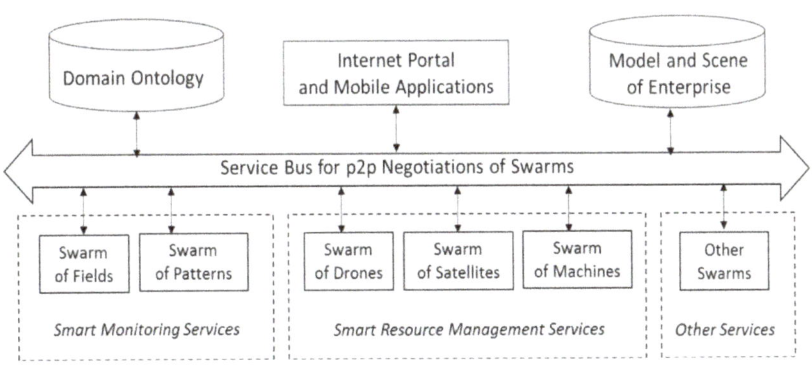

Figure 2: The architecture of Smart farming eco-system

The architecture will be open for entering new smart services, which will register itself in the knowledge base with the use of yellow pages for subscriptions.

Each smart service is designed as an autonomous multi-agent system having its own event queue, virtual world of agents with plan of deliveries, UI components, database, decision-making log, etc. Smart services can generate orders and negotiate deals with other services and users in real time. For example, when a new user enters the system and provides information on his fields, the smart service of satellites can analyze its coordinates and compare them with routes for satellites – determining the place and time of receiving images and proposing its service to the user. Then it can also trigger the service of image processing and pattern recognition, drone scheduling, etc.

For coordinated decision making, special protocols are developed for agents of service negotiations, which support the "round table" negotiations among experts (Fig. 3).

During the "round table" negotiations, all agents follow their basic interests; however, when a conflict is detected, they use special techniques to define and solve the conflict.

Let us consider the example for the following team of agents:

- Agent of satellite is identifying a pattern of illness on one of the fields;
- Agent of drone plans detailed imaging;
- Agent of insects and Agent of illnesses are competing for the detected filed;
- Agent of agronomy schedules a visit to the field;
- Agent of agronomy recognizes that there is not enough fertilizers;
- Agent of fertilizer is scheduling extra amount of fertilizers to the field;
- Machines and humans are scheduled for the operation;
- Agent of field requests control of executed operation, etc.

Thus, the key feature of the solution is to demonstrate "team work" of agents with cooperative and adaptive behavior of services. Examples of screens for crop monitoring are presented in Fig. 4.

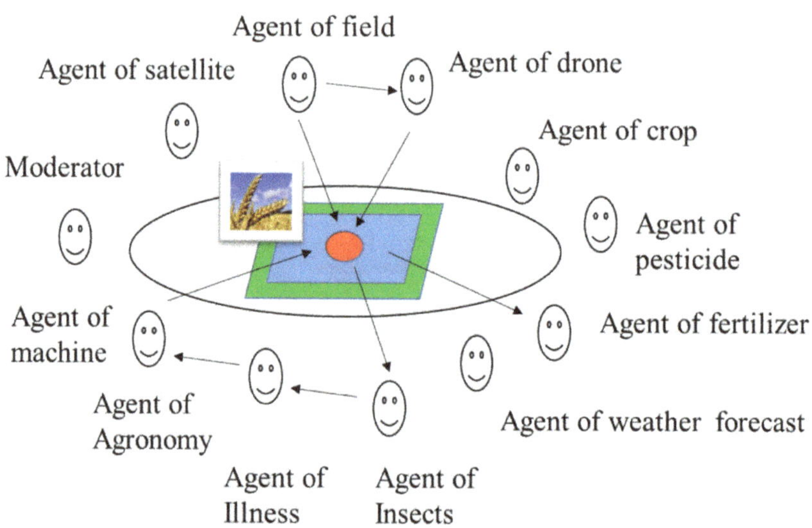

Figure 3: The Virtual Round Table for Coordinated Decision Making

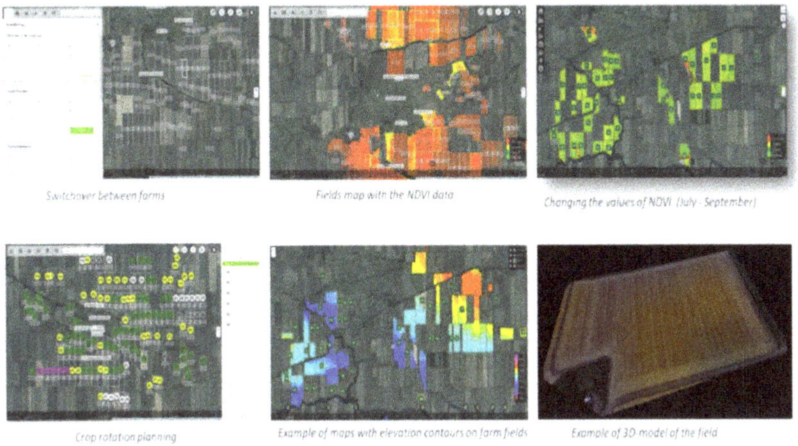

Figure 4: The examples of solution screens

Another most important part of the solution architecture is domain ontology of agriculture and knowledge base for the enterprise. Ontology as a method and tool of Semantic web for specification of domain-specific knowledge is widely discussed in literature [7–8]. In the agriculture domain there is also a number of R&D activities focused on formalizing different aspects of plant genetics, biology, etc. [9–10].

In our developments, we are focusing on integrating "parts" of agricultural knowledge, which makes the technology of precise farming applicable for different regions and closely connected with economics. The content of the knowledge base includes the following: types of soils, sorts of wheat, classes of machines and mechanisms, illnesses of plants, insects, pesticides, fertilizers, etc. The first priority is given to the knowledge that is battle-proven in regional practice and forms a list of precedents and success stories.

Ontology Editor and examples of domain ontology and knowledge base of enterprise are presented in Figs 5 and 6.

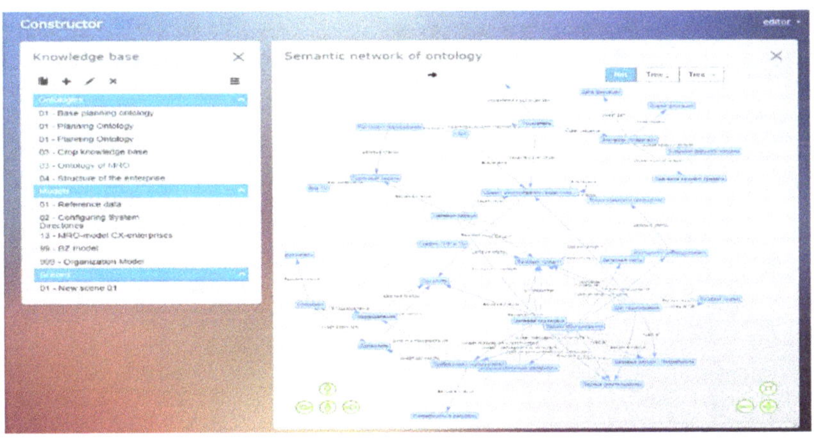

Figure 5: Ontology of agriculture

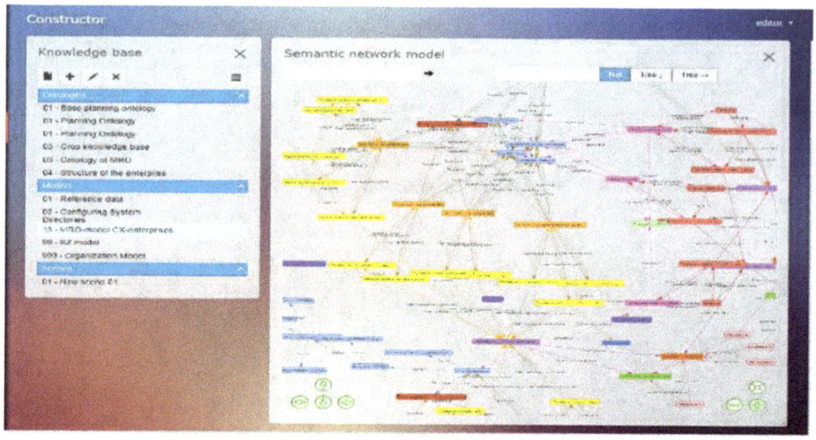

Figure 6: Ontological model of agricultural enterprise

The developed smart services for resource management are given in Figs 7 and 8.

The use of multi-agent technology together with ontologies makes it possible to allocate, schedule and optimize resources in real time taking into account the dynamics of events and situations that continuously change in practice.

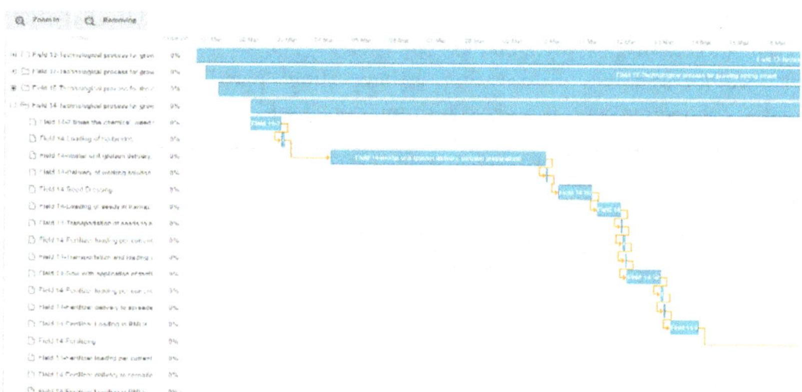

Figure 7: Example of machine operations scheduling

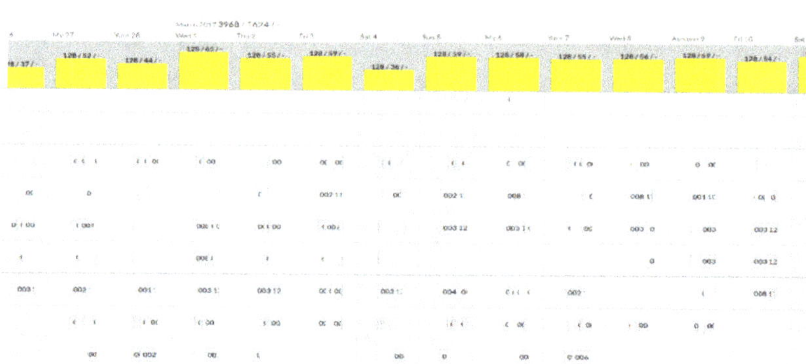

Figure 8: Dispatching daily operations

4 APPLICATION OF THE SMART FARMING SOLUTION

The current stage of research is focused on conceptual design and modelling of key functional and architectural decisions.

Today the first prototype of the solution is designed and is currently under testing for Peschanokopskaya Agro Group in Rostov region, which has wheat fields with more than 30 000 ha of surface fragmented in smaller sections. Agronomy specialists need to monitor all these fields covering large distances on a daily basis (more than 700 km a day).

The system can target at least 5–10% increase of profitability for agricultural business as well as a number of other benefits for farmers:

- increasing crop productivity while improving crop quality by eliminating soil degradation through operational monitoring of fields;
- reducing harmful influence on the environment and the Earth's biosphere by using methods of precise agriculture based on calculation of exact fertilizers using operative information about vegetation indices;
- increasing the full transparency of business and payback using advanced IT-systems and decision support, etc.

In the future, the solution can be applied for a wide spectrum of agricultural enterprises, including manufacturing of wheat and rye, rice and soya fields, grape and apple gardens, coffee and tea plantations, olive groves and cork tree gardens, etc.

5 CONCLUSIONS

In this paper, we have introduced the concept of digital platform and eco-system for precise agriculture, which is related to the theory of complex adaptive systems.

In the proposed Smart Farming solution, decision-making support for farmers is based on self-organization of agents representing smart services for monitoring, resource planning, pesticides, fertilizers, etc. The architecture of the proposed solution is designed as a digital knowledge-based eco-system of smart services for precise agriculture, which uses enterprise service bus and p2p communication of smart services. The virtual "round table" becomes the key mechanism for coordinated decision making of agents representing smart services. The use of multi-agent technology together with ontologies makes it possible to adaptively plan and optimize resources in real time, taking into account the uncertainty and dynamics of events.

The future works will be focused on developing ontology and knowledge base for precise agriculture of wheat and step-by-step implementation of smart services. It will also include detailed design of virtual "round table" protocols and implementation of autonomous smart services as well as methods and tools for learning from experience.

ACKNOWLEDGEMENTS

The work was supported by the Ministry of Education and Science of the Russian Federation in the framework of contract agreement No14.574.21.0183 – the unique identification number is RFMEFI57417X0183.

REFERENCES

[1] Sfiligo, E., Top 10 technologies in precision agriculture. September 9, 2016, available at https://www.therobotreport.com/top-10-technologies-in-precision-agriculture/

[2] Precise farming is an innovation in a system of resource-saving farming. The Federal Center for Agricultural Consulting and Retraining in the Agro-Industrial Complex, available at http://mcx-consult.ru/d/77622/d/tochnoe-zemledelie.pdf

[3] Nikolaev V.S. (ed.), Fundamentals of agricultural technology. Agriculture and plant growing [In Russian]. - Moscow: Bylina, 555 p, 2000, available at https://www.twirpx.com/file/1141268/

[4] Hopkinsm, M., 10 New Mobile Apps for Precision Agriculture, December 5, 2017, available at http://www.precisionag.com/service-providers/10-new-mobile-apps-for-precision-agriculture/?utm_source=PrecisionAg&utm_medium=website&utm_campaign=callout

[5] Digital Business Ecosystems & The Platform Economy, available at https://www.gartner.com/technology/topics/business-ecosystems.jsp.

[6] Rzevski, G. & Skobelev, P., *Managing Complexity*, WIT Press: Boston, 2014.

[7] Ontology Summit 2017 Communiqu´e – AI, Learning, Reasoning And Ontologies, available at https://s3.amazonaws.com/ontologforum/OntologySummit2017/Communique/OntologySummit2017Communique_v8.pdf

[8] Borgest, N.M. & Korovin, M.D., Ontologies: current state, overview [In Russian]. *Ontology of Designing*, **2**, pp. 49–55, 2013.

[9] Cooper, L., Meier, A., Laporte, M-A., Elser, J.A., Mungall, C., Sinn, B.T., Cavaliere, D., Carbon, S., Dunn, N.A. & Smith, B., The Planteome database: an integrated resource for reference ontologies, plant genomics and phenomics. *Nucleic Acids Research*, **46(D1)**, pp. D1168–D1180, 2017.
https://doi.org/10.1093/nar/gkx1152

[10] Repository for the Plant Ontology, available at https://github.com/Planteome/plant-ontology

REGENERATIVE DEVELOPMENT – BUILDING EVOLUTIVE CAPACITY FOR HEALTHY LIVING SYSTEMS

BRUNO DUARTE DIAS
CITAD, Lusíada University of Lisbon, Portugal.

ABSTRACT

Regenerative Development is a systemic approach that sees, for example, global challenges as the climate change hypotheses, as an opportunity to develop co-agency among several cultures and fields, toward a shared purpose that can reconcile the fragmented relationship of human and natural systems and multiple problems within a place. From this point of view, human systems (business, education, food and building industries, etc.) and natural systems (watersheds, rivers, forests, ocean, etc.) are aal seen as living systems. Projects can be designed and redesigned as change agents that have the capacity to influence and develop the vitality and capacity for evolution of these living systems as a whole. Resilience and sustainability become by-products of a strong and better structured whole system. This is a review paper that aims to uncover the patterns that emerge from different fields of human endeavours were the Regenerative approach is being applied such as business, supply chains, agriculture, building industry and others. The intention is to bring concrete examples that can reveal the differences between regenerative and conventional sustainable approaches.
Keywords: Evolutionary Systems Thinking, regeneration, regenerative development, sustainability.

1 INTRODUCTION

Sustainability is becoming a very ambiguous term. An article revealed that until 2007 there were an estimated three hundred definitions of 'sustainability' and 'sustainable development' exist broadly within the domain of environmental management and the associated disciplines which link with it, either directly or indirectly [1].

The architect Bill Reed claims that "Sustainability is not deliverable. Sustainability is not a thing. Sustainability is not simply about efficient technologies and techniques. It is about life – a process by which living things such as forests, neighbourhoods, people, businesses, mushrooms, and polar bears ensure their viability. It is a process of reciprocal relationship – a process by which living things support and are supported by a larger whole" [2]. For example a very high energy efficient building can't be considered sustainable, because it doesn't sustain life."

Although the good intentions of sustainability goals to reduce impact and increase efficiency and effectiveness on the use of resources, these goals are a poor response considering that human population and ecological footprint are increasing at an exponential rate [3]. Most sustainable solutions aim to solve problems, but usually when we solve a problem others emerge.

For example in 2015 the increased heat and drought in California cause dams to run out of water. The solution was to fill the surface with floating black balls to stop the evaporation. Instead of going to the source of the problem, this mechanical way of solving problems generates unintended consequences. For example, in 2017, after nine years, these balls had to be removed from a reservoir in Silver Lake, LA, because its water became undrinkable [4].

2 WHAT IS REGENERATION

In biology, regeneration is the process of renewal, restoration, and growth that, makes genomes, cells, organisms, and ecosystems resilient to natural fluctuations or events that cause disturbance or damage [5]. The biologist Richard J. Goss argued that regeneration is a primary

DOI: 10.2495/DNE-V13-N3-315-323

attribute of all living system, without which could be no life. To understand how nature sustains life, we need to move from biology to ecology, because sustained life is a property of an ecosystem rather than a single organism or species. Over billions of years of evolution, the Earth's ecosystems have evolved certain principles of organization to sustain the web of life [6].

2.1 Introduction to Living Systems

The concept explored in this paper is based in the perspective of a particular living systems thinking, school of thought. The dictionary of Etymology argue that Systems is a word that came from the Greek word "systema", that means an organized whole, a whole compounded of parts. Living system is here defined as a self-organized whole.

But as Pamela Mang argue – From a regenerative perspective "...natural systems and human systems, ... are all living systems" [7]. In the field of systems thinking, James Grier Miller defines living systems as a special subset of concrete systems. They can be composed of the moneres, protists, fungi, plants, animals, groups, organizations, societies, and supranational systems. Living systems are open systems that exchange matter-energy and information with its environment; they maintain a steady state of negentropy even trough entropic changes; they have more than a certain minimum degree of complexity; descended from some primordial DNA (deoxyribonucleic acid) common to all life, or have a charter; they are constitute of organic compounds, and in some cases, nonliving components; they have a decider system that enable the interaction of components and subsystems; they subsystems are integrated together to form actively self-regulating, developing, unitary systems with purposes and goals [8].

2.2 Regeneration in the context of sustainability

In the context of the field of sustainability, the term regenerative emerged for the first time connected with agriculture practice, in the paper submitted by the Rodale Institute with the title – Potential for regenerative agriculture in developing world [9]. In 1994 John Tillman Lyle introduce the term "regenerative design" as an approach to the design of urban landscapes which enables them to regenerate lost ecosystems [10].

Most approaches to sustainability focus on social and environmental health using traditional reductionist to solve problems by applying best practices. Regeneration aim to building healthy human and natural networks, drawing on universal principles and patterns, in which "sustainability" and resilience is a byproduct of systemic health [11]. Is not just about reducing impact or doing good things to nature, it is about recognizing that we are part of nature and that our activities need to be designed and integrated in a way it contribute and add to the whole, instead of just extract.

Regeneration is different form restoration, because life evolves and is not static, we can never restore a living system to its original condition. When we speak of restoration (of a woodland, a riparian system, a wetland) we are speaking of restoring a system's capability to continuously self-organize and evolve [12]. Regeneration is a concept that aims to develop will and invite people to take care of our life support system (earth systems).

2.3 Three possible aims of regeneration

A regenerative project or initiative always comes from an organization (company, non-governmental organization, etc.). A regenerative organization is authentic, is purposeful, it really wants to exist to make their communities (human and natural), more viable and vital.

1. **Regenerative solutions (inside the limits of propriety)** – are related to concrete solutions that are applied to agriculture, agroforest, landscape, urban design, etc. When applied in context with the unique characteristics of a place, these techniques can improve the quality of the soil, retention of water, increase of biodiversity, like for example regenerative agriculture. These are great and necessary solutions, but if we design only from a solution point of view we can still design our projects as islands close to limits of the project or propriety.
2. **Systemic Regeneration of place (around the project)** – A project can be seen as a dynamic piece that will move the system in a way it can regenerate our degenerate several other systems to which it is connected. Before starting a design process is necessary to understand the existing web of relationships and see how the project can work with them. If well designed and implemented, a project or initiative can contribute to increasing the vitality and capacity for evolution of human and natural systems.
3. **Systemic Regeneration of field of endeavour (in the world)** – At this level, an organization may aim to push or regenerate how a particular field of endeavour is done in the world. For example, the Regenesis Group is an organization that is "dedicated to regenerating the way human beings develop and inhabit ecosystems".

Several projects work on just one of the above three aims. Others, like the Playa Viva example, work on all three together.

3 REGENERATIVE DEVELOPMENT

This concept was proposed in 1995 by the Regenesis Group a North American collaborative, composed of a multidisciplinary team. The practice process of Regenerative Development is based on the Living System School of Thought originally developed by Charles Krone, to enable thinking about organizations as living systems – what organizes and orders them, how they are structured, how they evolve, etc. It uses systemic frameworks and developmental processes to consciously improve the capacity to apply systems thinking to the evolution of human or social and natural living systems [13].

In the framework levels of work developed by Krone, Regenerate or Regeneration is a level of work that addresses the unrealized potential inherent in the relationship between a given system and the larger systems within which it is nested [2].

To "regenerate" include three key ideas: a radical change for the better; creation of a new spirit; returning energy to the source [14]. The organization working from evolutionary systems thinking sees its primary purpose as regenerating the field "they/we" all exist in and seeking to harmonize toward where the field seems to be unfolding [15].

3.1 Main actors on the field of Regenerative Development

According to Raymond Cole, and other main actors on the field of sustainability, Regenesis Group has provided by far the most significant contribution to defining the scope, emphasis and key principles of regenerative development [16]. The Regenesis school of thought has also inspired the development of a number of tools, such as REGEN, a regenerative design framework developed by architectural firm Berkebile Nelson Immenschuh McDowell (BNIM) for the US Green Building Council; and LENSES, a facilitation tool developed by the Institute for the Built Environment at Colorado State University [7].

This paper will consider the definitions provide by the Regenesis Group to continue exploring what Regenerative Development aim to accomplish and explore some of the thinking that is behind. In this paper the implications of Regenerative Design are not going to be addresses superficially, although they are an important part of the Regenesis approach.

Regenerative development is define by Regenesis as "A system of technologies and strategies for generating the patterned whole-system understanding of a place, and developing the strategic systemic thinking capacities and the self-organizing and self-evolving stakeholder engagement/commitment required to ensure regenerative design processes to achieve maximal systemic leverage and support" [17].

Sarah Jenkin and Maibritt Pedersen Zari write that regenerative development "investigates how humans can participate in ecosystems through development, to create optimum health for both human communities (physically, psychologically, socially, culturally and economically) and other living organisms and systems" [18].

3.2 Regenerative development in practice

It seems that Regenesis makes the translation from theory to practice of Fridjot Capra statement: "the living world evolution cannot be limited to the adaptation of organisms to their environment, because the environment itself is shaped by a network of living systems capable of adaptation and creativity. They co-evolve. The evolution of living organisms is connected with the evolution of their environment that together they constitute a single evolutionary process." The shift is from evolution to co-evolution [19].

In the book Regenerative Design and Development, the authors argue that in order to comprehend regenerative development is necessary to introduce three key ideas:

Regeneration as enabler of evolution – Regeneration enables living systems to evolve by expressing their latent **potential** in the form of new value in the world. In this way, what exists now can move toward what could be in the future.

Working in place – The idea of place is a way for people to envision the unity of humans and natural systems in a concrete and specific way. In each place on Earth, natural and cultural systems express themselves uniquely (although often the qualities that differentiate them are masked by the effects of media and the global economy). This means that if we wish to engage in co-evolutionary partnerships with nature, we have to do so place by place, discovering opportunities and solutions that are indigenous to specific locations rather than generic to everywhere.

Developmental processes – A regenerative development project leaves behind more than physical structures; it does more than benefit the surrounding natural and social communities. It also grows new capability and capacity in the people that it touches. This is accomplished by including human development in every aspect of a project [2].

3.3 Regenerative development criteria

Table 1 expresses the criteria to be considered in the work of regenerative development. This content was taken of a presentation that Ben Haggard made in the Commonwealth workshop in 2017 [20].

This table illustrates that to intervene in complex living systems it is necessary to move from a reductionist approach to a holistic one. It's important to underline that problem solving, standard and best practices are appropriated to address complicated systems (like cars or planes). But when applied to Living Systems a problem solving approach generates inefficient and ineffective solutions/designs that may generate degenerative change in the whole system.

The following definitions were draw from a document presented in the Regenerative Summit and event organized by Carol Sanford. She is a regenerative practitioner with several

Table 1: [Seven criteria of regenerative development – Ben Haggard, presentation to RDRCC, May-18th-2017]

Moves Away From	Moves Toward
Parts and fragments	Living wholes
Commonality and averages	Essence, fostering differentiation
Problems and ideals	Inherent, specific potential
Quid pro quo and transactional	Systemic reciprocity among all life
Isolated and independent entities	Nested Systems
Shotgun approaches	Nodal Interventions for systemic effects
Command and control	Development of capacities & capabilities

decades of experience, using exactly the same Living Systems School of Thought, but applied to the field of economy, business, entrepreneurship, etc.

Carol Sanford argues that, Regeneration is found only in the world of living systems. It offers us a paradigm for understanding life. Machine and other non-living metaphors cause us to conceive inaccurately. Living Systems offer us a set of First criteria that allow us to understand the world as alive and "at work". A regenerative approach can be accessed through these seven foundational Criteria [21].

First criteria

Whole – systems, businesses, industries are living wholes. We tend to work on the parts as problems with considering the effects on the whole that we want to regenerate.

Essence – By honour and work from the singular characteristics of a being (person, company or place) we are able to discern the conditions in which a being is likely to thrive.

Potential – instead of looking to solve problems by applying best practices or imported ideas, projects are designed from the potential, or what could be but is not yet manifested.

Reciprocity – reciprocity aim to establish relationships of mutual benefit that benefit the system health with care for contributions and outcomes for all.

Nested Systems – Embedded within greater and lesser systems, each playing a core role in the success of the whole and other nested wholes.

Nodal – Seeking points for intervention that evoke systemic beneficial effects that have significance to the system. Similar to acupuncture, where a single point or set of points are recognized as most effective for systemic regeneration.

Development – Instead of trying to control, regeneration seeks to grow and develop capacity and capability in each and all entities to realize their essence.

Regeneration is not about technology or solutions. Regeneration is a way of thinking and a way of being in the world. Solutions are part of the equation but they only start to become concrete in a late phase of the design process.

4 TWO EXAMPLES OF APPLY REGENERATIVE DEVELOPMENT FRAMEWORKS

4.1 Regenerative development to reverse climate change

The Commonwealth and the Cloudburst Foundation organized the Workshop "Regenerative Development to Reverse Climate Change" in London from 28 to 29 November 2016 to launch a process to develop a viable, comprehensive and effective program for regenerative development. From around the world, a group of over 50 leading experts from multiple fields came together to share and discuss ideas on policy, technology and best practice. This notion of cooperation or collaboration fosters another way of seeing sustainability, which is grounded in the mind-set to reduce or mitigate the impacts we generate.

Hopefully, Regenerative Development has the potential to realize and exceed the Paris Agreement on climate change, and the Sustainable Development Goals to end poverty and hunger, boost economic growth, reduce inequality and maintain peace and stability. It is a re-thinking of the way we live with our land, forests, cities and oceans. It is based on the idea that we can rebuild ecological health, socio-cultural vitality, economic value and human potential in an integrated and reciprocal way [22].

As Ben Haggard argue "The technologies and strategies we are talking about need to be undertaken in consultation with local people. This is key to the idea of regeneration. What we are really up to here is building new capacity into every community, nation and ecosystem where these strategies are applied. So it is not just the strategies, it's also the capacity that is being built into the systems surrounding those strategies. This is how reversing climate change also becomes an engine for social and economic regeneration" [22].

Case study conclusion

Regenerative development is an integrative and inclusive concept by nature. It has the ability to look at the uniqueness of a situation or project and merge other tools and concepts, not because is on a check list, but because they are appropriated and can add value in a particular context.

4.2 Example – Playa Viva eco-resort

This eco-resort was implemented in a 86 hectares plot located in Zihiuatanejo, Baja California, Mexico. The master plan includes 60 Casitas, 13 single lots, a town square and a boutique hotel.

Initially, with a sustainable mind set, the main goals were to produce all the energy on site from renewable sources, use eco-friendly building materials and local building techniques.

After engaged in a systemic regenerative thinking approach the master plan of the resort took a different direction [2].

Regenesis developed a tree part process that is flexible and adaptable to the particularities of each project. This case study will be framed in this three part process although in reality the process is much more complex and organic.

Research Phase – Understand place

Before the design process begin is important to "understand the site and its community in relationship to place, through the dynamics, patterns, relationships and potential inherent in that specific place; and to conceptualize how, through right relationship, the project can be a regenerative force" [7].

Some of the things that emerged were: that property and the adjacent land were ecologically degraded after 80 years of intensive palm oil farming. The young population was leaving this rural area for lack of work. The resort beach was a turtle nesting area. Local people dug up, sold and consumed turtle eggs (illegally) which lead to the decline of the turtle population. One of the slopes on the property was the remaining structure of a pre-Columbian pyramid. That area had been a trading centre of the Aztec empire.

After having a notion of the particularities of the whole system, gradually began to realize how could integrate the project of the resort in this natural and human system [23].

Design phase – Designing for harmony

The aim is to harmonize the buildings or infrastructure with the patterns of the land and its cultural and ecological systems, to improve all, including the inhabitants of the place [24].

The design team and the developer try to understand what could be the role of the project in this place, that lead to the development of the master plan and several programs and initiatives:

The "Viva Praia Reserve" – a program was started with the aim of regenerating 73 hectares of the property, the coastal forest and the mangrove. They built a nursery of native and ornamental plants, controlled invasive species such as acacia, cleaning and restoration of rivers and streams, and improved water quality so that shrimp and other native bird species returned to the area [13].

Living beach Turtle Sanctuary – After understanding the importance of turtles to the ecological system, a program was created that was of local interest. It was made a maternity of turtles, the turtle hunters were hired by the resort to unearth and incubate the turtles. These program is an intersection of tourism and conservation, the aim is to increase the turtle population and also transformed the hunters in protectors [25].

Relationship with the community – Instead of the resort being thought to be an entertainment island, closed to the local community, the resort become the support for local development promoting social, natural and economic vitality. At one point local authorities offered the developer the possibility of make a direct access from the motorway that passed outside the local village. An offer that the developer refused because he realized that the community and the visitors could benefit each other from their passage inside the village [7].

Operational phase – Co-evolution

A regenerative project aims to set conditions in place which will ensure ongoing co-evolution and mutually beneficial integration of human and natural systems [25].

Playa viva resort opened in November 2009 using a living systems framework of adaptive, co-evolutionary growth. It made considerable changes at the economic environmental and social development of the community. The aim was to continue to develop the community in sustainability education, like permaculture design and recycle workshops. A Co-op of salt emerged from the interaction of the resort and local producers, to ensure quality and fair trade. The resort started some actions with the local community. Together with the local schools they set up cleaning days to keep the village, forest and rivers cleaner, and explained the benefits of waste separation. Do agriculture actions and plant trees outside and inside the resort. They supported and provided school and medical supplies to the population, the resort offered organic farming programs to locals, who subsequently began to provide organic food to the hotel and local markets [7].

Transformative experience for visitors – In addition to the experience of releasing turtles, tourists have the possibility of contacting stakeholders who regenerate natural systems, participate in local ecological projects, and participate in initiatives within schools.

Regenerative resorts – this is the latest project lead by David Leventhal (Playa viva owner). This project aims to build a collaborative of resorts that can expand the impact investment community. This illustrates the Systemic Regeneration of the field of endeavour, this project aim to make significant changes on what can be the role of a resort in a particular place [26].

Case study conclusion

There is a long list of projects and programs that are being develop by the Resort. What is present here is the culture of, what can we do more? How can we evolve the capacity of the systems that we are interdependent with?

Even the concept of success was redefined. Bill Reed one of the element of the project team stated that "the greatest success of this project can be seen by the return of young people to their village, increased biodiversity, economic and social vitality of the village, changing farming practices, egg hunters who became guardians of the turtles, etc." In another words the success is measured outside the limits of the project.

Instead of this project being thought to extract value from visitors and local communities the focus is to generate value in local human and natural systems, knowing that this will contribute to the viability of the Resort in such a way that it has become vital and nondispensable for the area.

Playa viva is a purposeful business that acts as a change agent were social responsibility and innovation are weave together and connected with the well-being of social and natural communities.

5 CONCLUSION

Only a few decades ago "civilized man" started to understand the complexity of the world and the impact that human activities have on the systems that support life. Until now we develop a extractive and passive mind-set, we take what we can without giving back. To make real changes our role must shift to become active in developing relationships of mutual reciprocal benefits between human and natural communities.

The practice of Regenerative development aims to take people and their endeavours from their assigned boxes, to open the possibilities of what a project or business can do for all the stakeholders involved.

Regenerative development is a change process that aims to reconcile and harmonize how human/social systems work with natural systems. It's a systemic approach that aims to build in projects, business and communities the capacity to regenerate the whole system. Any project or business can be the entry point to address the health of the whole system, if people can leave there assign and unassigned list boxes and open to the possibilities of the role that each person and organization can bring to place that inhabit.

ACKNOWLEDGEMENTS

This work was made at CITAD – Research Centre for Territory Architecture and Design – funded by National Funds through FCT – Foundation for Science and Technology under the Project UID/AUR/04026/2013

REFERENCES

[1] Johnston, P., Everard, M., Santillo, D. & Robèrt, K.-H., Reclaiming the definition of sustainability. *Environmental Science and Pollution Research – International*, **14**(1), pp. 60–66, 2007.
https://doi.org/10.1065/espr2007.01.375

[2] Mang, P. & Haggard, B., *Regenerative Development and Design – A Framework for Evolving Sustainability*, Wiley & Sons: Hoboken, New Jersey, 2016.

[3] Dias, B.D, Beyond Sustainability – Biophilic and Regenerative Design in Architecture, *European Scientific Journal*, 2015.

[4] CURBED, Ivanhoe Reservoir ' s shade balls are going away very soon, available at https://la.curbed.com/2017/4/4/15182772/ivanhoe-reservoirs-shade-balls-removed (Accessed 10 March 2018).

[5] Birbrair, A., Zhang, T., Wang, Z-M., Messi, M.L., Enikolopov,G.N., Mintz, K. & Delbono, O., Role of pericytes in skeletal muscle regeneration and fat accumulation. *Journal of Neurotrauma*, **4126**, pp. 1–65, 2015.

[6] Capra, F., Ecoliteracy in your Inbox!, available at https://www.ecoliteracy.org/article/new-facts-life (accessed 03 March 2018).

[7] Hes, D. & Du Plessis, C., *Designing for Hope*. Routledge: Abingdon, 2014.

[8] Miller, J.G., James Grier Miller Living Systems . *Panarchy.Org*, pp. 1–37, 1978.

[9] Terra Genesis International, "Levels of Regenerative Agriculture," September, 2016.

[10] Doyon, A. & Hes, D., Positive buildings, an opportunity to increase resilience, 2014.

[11] Fullerton, J., How Universal Principles And Patterns, 2015.

[12] 7group and Reed, B., *The Integrative Design Guide to Green Building - Redefining the Practice of Sustainability*, John Wiley & Sons: Hoboken, 2009.

[13] Mang, P. & Reed, B., Designing from place: a regenerative framework and methodology. *Building Research & Information*, **40**(1), pp. 23–38, 2012. https://doi.org/10.1080/09613218.2012.621341

[14] Mang, P. & Regenesis Group, "Regenerative design and the evolution of the sustainable design field," 2001.

[15] Sanford, C., A theory and practice system of ' systems thinking ': with an executive ' s story of the power of ' developmental ' and ' evolutionary ' systems thinking. *Why a New Typology of Systems Thinking ?*, pp. 1–23, 2004.

[16] Cole, T.R.J. & Oliver, A., The Next Regeneration, 2012.

[17] Mang, P. & Reed, B., Regenerative development and design. *Encyclopedia of Sustainability Science and Technology*, **I**, pp. 8855–8879, 2012. https://doi.org/10.1007/978-1-4419-0851-3_303

[18] Zarim, M.P. & Jenkin, S., "Degenerative Development," no. 1, pp. 1–10, 2009.

[19] Capra, F., *The Web of Life: A New Scientific Understanding of Living Systems*. New York: Anchor Books, 1996.

[20] The Commonwealth, "The Commonwealth and Cloudburst ' s Workshop on Regenerative Development to Reverse Climate Change Table of Contents," 2016.

[21] Sanford, C., What Is Regeneration?, 2017 available at https://theregenerativebusinesssummit.com/regeneration/ (accessed 20 February 2018).

[22] The Commonwealth, "Regenerative Development to Reverse Climate Change Workshop," 2017.

[23] Group, R., "Playa Viva — More Than a Sustainable Development." available at http://regenerationalliance.com/projects/playa-viva/ (accessed 30 March 2018).

[24] Stay, B.Y., Contest, P., Resort, E., Hotel, E., Tourism, S., Eco, F., Lodge, L., Retreat, E., Kahn, S. & Leventhal, D., History : The Playa Viva Story History of Place : Archaeological Site, 2017, available at http://www.playaviva.com/en-gb

[25] Luna, M., Taking a hard look at recent developments. *Director*, available at https://playaviva.wordpress.com/2017/08/28/taking-a-hard-look-at-sustainable-seafood/ (accessed 05 March 2018).

[26] Leventhal, D., "VALUES WHO WE ARE," 2018, available at www.regenerativeresorts.com (accessed 30 March 2018).

A ROUGH SET BASED ALGEBRAIC APPROACH TO MODELLING COMPLEX SYSTEMS

D. SITNIKOV[1], O. RYABOV[2], I. MISHCHERIAKOV[1] & A. KOVALENKO[1]
[1] Kharkiv National University of Radio Electronics, Ukraine.
[2] National Institute of Advanced Industrial Science and Technology, Japan.

ABSTRACT

"A complex system is considered as an algebraic structure having specific properties that do not allow expressing precisely the meaning of information objects included in the system. An algebraic definition of complexity has been given. The complexity has been considered from two viewpoints. A system can be considered to be a complex one if (a) boundary regions for system's objects and processes are broad; (b) possibilities for system's decompositions are limited. Difficulties associated with complex system description and decomposition have been discussed in the framework of rough set methodology. A method for extracting salient features of information objects available in the system has been outlined. Some theoretical aspects of rough set based analysis of complex systems have been discussed. All the operations designated to obtaining new knowledge or data patterns in complex systems are algebraically described from the viewpoint of algebraic systems including predicates (in particular, finite ones) and operations on them. Thus, a system is considered to be complex if there is a great degree of uncertainty in the data and/or there are some serious problems with system's decomposition."

Keywords: big data, complex system, knowledge discovery, rough set, system decomposition, uncertainty in data

1 INTRODUCTION

Now there are no universal formal tools allowing us to describe and treat any kinds of complex systems. Nevertheless, we need to have some instruments and semantics to deal with systems complexity. Therefore, we try to consider the concept of a complex system from a logic-algebraic viewpoint, thereby considering it as a certain algebraic structure with predicates describing the available knowledge about the system and operations over them allowing us to describe some pieces of information in the system via the available knowledge structures. From our viewpoint, one of the most important features of a complex system is some level of uncertainty in data that does not allow making precise decisions based on the available information. In our opinion, the complexity lies not necessarily in a complicated database or knowledge base structure and not in a complicated and interconnected information flows. Also, the complexity does not necessarily mean many interconnected nodes in the system. We think that what does matter is the following: (a) the level of uncertainty we encounter when making decisions related to the system objects and processes; (b) the possibility of decomposing the system with losing, of course, some information on its elements. These two are associated with each other as the level of uncertainty obviously increases when we try to reduce the system to some subsystems and relations between them. Conversely, if system objects and processes become more vague and difficult to describe and deal with, the system's decomposition to subsystems and relations between them becomes more difficult. These basic ideas have led us to an understanding that mathematical instruments dealing with vagueness and uncertainty in information systems should come into play. We have chosen rough set theory as the starting point for the following reasons. Other theories like fuzzy set theory or theory of evidence do allow treating uncertainty in data but do not allow convenient algebraic operations with information objects, which are necessary for understanding basic interrela-

DOI: 10.2495/DNE-V13-N3-324-329

tions between them. Really, even the classical topological approach of Z. Pawlak extends the possibilities of probabilistic and statistical approaches. As we demonstrate in our papers, from our viewpoint, the opportunities of approximation based reasoning are far broader than Prof Pawlak supposed at the very beginning when creating his purely topological approach.

The founder of rough set theory Zdzislaw Pawlak considers his approach to modelling uncertainty and vagueness in data as very much promising for the future of intelligent industrial applications [1]. In his opinion confirmed by top quality research and practical results that have been obtained during the last decades, rough set models and tools are particularly suited for solving real-world problems in medicine, market and financial analysis, engineering, banking, pharmacology and other fields. Prof. Pawlak and his followers believe that rough set theory can be effectively used for modelling industrial processes and solving real-life problems in the following areas: material sciences, intelligent control, decision support systems, machine diagnosis, and neural networks.

From the viewpoint of prof. Pawlak, rough set methodology allows data mining and knowledge discovery by identifying relationships that cannot be found with the help of purely statistical approaches. Pawlak says that rough set based data reduction and assessing the importance of particular data seem to be very much useful in many branches of intelligent applications as well as rough set based generating logic rules that are easy to understand and interpret. We completely agree with this point of view.

Moreover, the approach outlined in the next section shows that the basic rough set ideology formulated by Prof Pawlak allows operating with Big Data, which are in principle heterogenous and multivariate. We do not argue with Prof Pawlak but follow his main ideas. These ideas have led us to some principally new approaches to treating uncertainty in data. What we suggest is this: (a) start with the available data structures and allowable operations on them; (b) describe any objects in a complex system via the well-known ones with the help of the available operations.

2 ALGEBRAIC APPROACH VERSUS KNOWLEDGE GRANULARITY BASED APPROACH

Almost everything that is related to rough set theory is based on so-called indiscernibility relations that induce some kinds of knowledge granularity, which allows classifying objects into similarity classes containing elements that cannot be discerned with respect to specific features. Of course, there is an obvious advantage of such a topological approach to treating uncertainty in data. As in the case of the classical database theory, the classical rough set theory by Z. Pawlak gives us a lot of opportunities to structure vagueness in our knowledge about the real world. In the case of rough sets, the indiscernibility relation and subsequent knowledge granularity allow bringing some order to data uncertainty, the possibility of operating with some basic "bricks" of the available information.

Nevertheless, the topological approach to treating vagueness in data implies some essential limitations as to our understanding of the uncertainty concept. When we encounter data points that cannot be discerned with respect to some features, we are uncertain about a class (cluster) to which different points should be attributed. This is how we understand the uncertainty concept following Pawlak's approach.

In our series of papers [2–5] we try to give the rough set ideology a bit different perspective by avoiding the usage of the knowledge granularity concept. Whereas the classical rough set theory is based on the indiscernibility relation (which can be equivalence, tolerance or other types of relations), we do not use such a notion and try to consider uncertainty from a logic-algebraic viewpoint. Then a natural question arises: why do we think that we follow the rough set approach, if we do not accept the very first step in data analysis, namely the consideration of knowledge indiscernibility and granularity? The answer is this. We do not start with granularity. We start with rough approximations right away. We suppose that we originally have something (some knowledge about the world) expressed in terms of information objects, under which we understand not only data points but any predicates that can be defined on these elements (for example, on database records). It seems quite natural as any knowledge can be mathematically represented in the form of relations (predicates) or functions (that can be considered as a subtype of relations).

We introduce the concept of an approximation language, i.e. predicate operations allowed for describing dependencies between information objects. We describe information objects in terms of other objects as exactly as it is allowed by the approximation language. Thus the basic idea of Pawlak about approximation based reasoning stays intact. Moreover, it seems quite natural to start with approximating something that cannot be expressed exactly in terms of the available semantics. Following such an approach we do not need to suppose a priori which objects can be discerned with respect to some specific relations. All we need is just to find a logic-algebraic function that approximates the object under consideration "from below" and "from above". An object under consideration can be any predicate, and according to our approach it should be expressed in terms of the other predicates available at the moment with the help of the predicate operations included in the approximation language.

The classical topological approach implies that there are the only upper and lower approximations for a rough set. We are not limited to a single approximation, but if an approximation cannot be improved in terms of the approximation language, we call it an exact approximation. According to what has been done before, we can now present a general definition of reasoning based on our rough set algebraic methodology, and we hope that industry professionals from various fields will be able to apply it to solve their specific knowledge discovery and machine learning problems outlined by Z.Pawlak and his followers.

1. Let $P_1(x_1,x_2,...,x_m), P_2(x_1,x_2,...,x_m),...,P_n(x_1,x_2,...,x_m)$ be predicates, i.e. functions taking on values 0 or 1. We interpret them as the accessible knowledge structure.
2. Let $Q_1(P_1,P_2,...,P_n), Q_2(P_1,P_2,...,P_n),...,Q_k(P_1,P_2,...,P_n)$ be operations over predicates, the set of which we call the approximation language. This language allows expressing information objects in terms of the accessible knowledge.
3. Let $X(x_1,x_2,...,x_m)$ be a predicate that we call an information object, which is required to be described in terms of the predicates $P_1,P_2,...,P_n$ with the help of the approximation language.
4. Any solution F of the functional equation $X(x_1,x_2,...,x_m) \to F(P_1,P_2,...,P_n) = 1$, where F is constructed with the help of the approximation language, is called an upper approximation for X.
5. If a solution F^* of the above equation cannot be improved in the sense that any other solution F satisfies the equation $F^* \to F = 1$, then we call F^* the exact upper approximation for X.
6. Any solution F of the functional equation $F(P_1,P_2,...,P_n) \to X(x_1,x_2,...,x_m) = 1$, where F is constructed with the help of the approximation language, is called a lower approximation for X.
7. If a solution F_* of the above equation cannot be improved in the sense that any other solution F satisfies the equation $F \to F_* = 1$, then we call F_* the exact lower approximation for X.
8. The function $F^* \wedge \overline{F_*}$ is called the boundary region for the object X.

The boundary region plays an important part in the classical Pawlak's rough set theory. It is a domain in which we are not certain to which set (class, cluster) an element should be attributed. If in the classical theory the boundary region is just a set of elements that we do not know exactly where to place, in our generalizations the boundary region is an object of a higher order. In the above model it is a function over predicates. This function along with the possibility of decomposing the system is suggested to be a measure of system's complexity. It should be noted that in case this function takes on only zero values, the object X under consideration can be precisely described in terms of the available objects with the help of the approximation language. If the boundary region represents a function equaling 1, it means that nothing can be said about the object under consideration in terms of the available knowledge. This is why we suggest this function as a measure (not the only one) of the "algebraic complexity" of a system.

It can be easily shown that the main properties of approximations in Pawlak's theory hold, but the above algebraic definitions allow us to generalize the classical Pawlak's concepts in different directions by variating approximation languages and the number of the original variables. For example, we can consider binary or ternary predicates and quantifiers as well as other operations on predicates.

The suggested approach allows explaining the meaning of an information object in terms of the accessible knowledge represented by predicates and allowable predicate operations. It is supposed that the predicates

$$P_1(x_1,x_2,...,x_m), P_2(x_1,x_2,...,x_m),..., P_n(x_1,x_2,...,x_m)$$

are obtained as a result of experiments and human intuition. As soon as these predicates and the necessary approximation language (semantics) are defined, the reasoning is carried out with the help of upper and lower approximations. Again, following the classical rough set theory approach, the exact upper approximation means a possible dependence, and the exact lower approximation means a necessary dependence. It should be noted that not only the exact approximations may be interesting for the investigation of hidden dependencies in data, but also other approximations may be useful. This point needs further research. We have shown that the exact approximations can be easily obtained in the case of unary predicates and Boolean operations [2], but in more complicated cases (binary predicates, quantifiers etc.) it may turn out that the exact approximations are difficult to calculate, and we can impose some limitations allowing us to select approximations close to the exact ones.

The boundary region in our understanding plays even a greater role in the sense that it allows extracting salient features from the available knowledge bases. The whole scheme of obtaining the importance of features is simple. We (a) eliminate a feature (a predicate in the general case) from consideration; (b) recalculate the upper and lower approximations *without* this predicate; (c) estimate how much the boundary region has changed as a result of this elimination. If the boundary region has changed very much (it can be measured), the eliminated feature is important. If the boundary region has changed a little bit, the feature is non-salient.

3 DECOMPOSITION OF A SYSTEM. MEASURE OF COMPLEXITY

As we mentioned in Introduction, not only the *uncertainty factor* but also difficulties related to decomposing the system are extremely important for defining the system as a complex one. In this connection we should note that describing a certain predicate in terms of some predicates depending on *sets of variables that do not intersect*, represents a degree of complexity associated with the *decomposition capacity factor*.

Consider a simple example. We should describe the predicate $X(x_1, x_2, ..., x_m)$ via some other predicates depending on the same variables. Suppose that we can find a predicate P "close" to X ("closeness" is a matter for a special discussion) such that

$P(x_1, x_2, ..., x_m) \rightarrow X(x_1, x_2, ..., x_m)$ or $X(x_1, x_2, ..., x_m) \rightarrow P(x_1, x_2, ..., x_m)$. If the predicate P can be decomposed to predicates of smaller dimensions (ideally $P = P_1(x_1)$ & $P_2(x_2)$ &....& $P_m(x_m)$, although the conjunction between these predicates is not the only option; we can consider any formula constructed with the help of the approximation language operations) so that the available knowledge is not lost but just decomposed, it means that the system can be simplified. If it is not possible, the system stays complex. The main idea is this. The harder the decomposition of the upper and lower approximations, the more complex the system is.

4 CONCLUSIONS

In this short paper we have outlined a mathematical (and, may be, philosophical) approach to understanding the complexity of an information system. We considered a general method of treating uncertainty in data, which is alternative to the classical rough set theory methodology, but it leads to approximation based reasoning with more general assumptions. As a rule, real-world intelligent applications have to deal with complicated data structures and Big Data. It means that a broader view on the rough set ideology can facilitate reasoning with elements of vagueness and uncertainty in the available knowledge. In our approach we do not require the existence of an indiscernibility relation in data, but we operate with the existing knowledge structure represented by sets of predicates and allowable predicate operations. We consider the complexity problem from two interrelated viewpoints: (a) how precisely we can describe an information object via the available and understandable objects; (b) how the obtained descriptions can be simplified by decomposing the available predicates. In our opinion the above two factors tell us a lot about the system complexity.

It should be noted that our approach cannot be considered as a universal one, although it is quite general. We have not considered here a lot of important issues: emergence, self-organization, adaptation etc. We have just concentrated on the following two aspects: uncertainty (this factor surely plays an important part in decision-making process in complex information systems); possibility of efficient decomposition for an information system objects and processes (in case of serious problems with such a decomposition the system should be recognized as a complex one). Even these matters have been considered in the general algebraic systems framework. Therefore, we hope that future research into these matters will be carried out and allow forming a strong general algebraic theory of complex systems.

REFERENCES

[1] Pawlak, Z., AI and intelligent industrial applications: the rough set perspective. *Cybernetics and Systems*, **31**, pp. 227–252, 2000. https://doi.org/10.1080/019697200124801

[2] Sitnikov, D. & Ryabov, O., An algebraic approach to defining rough set approximations and generating logic rules. *Data Mining V: Proc. 5th International Conference,* Malaga, Spain, 2004, pp. 179–188, 2004.

[3] Sitnikov, D., Ryabov, O., Titova, E. & Romanenko, O., A generalized algebraic approach to finding rough set approximations and generating logic rules. *Data Mining VIII: Proc. 8th International Conference,* Great Britain, pp. 3–12, 2007.
[4] Sitnikov, D., Titova, E., Ryabov, O. & Romanenko, O., An approach to finding reduced sets of information features describing discrete objects based on rough sets theory. *Data Mining IX: Proc. 9th International Conference,* Spain, pp. 3–11, 2008.
[5] Sitnikov, D., Titova, E. & Ryabov, O., A method for finding minimal sets of features adequately describing discrete information objects. *Data Mining X: Proc. 10th International Conference,* Greece, pp. 22–30, 2009.

Author index

www.ingramcontent.com/pod-product-compliance
Lightning Source LLC
LaVergne TN
LVHW062240200926
848484LV00006B/1579

* 9 7 8 1 7 8 4 6 6 3 6 7 4 *